Progress in Molecular and Subcellular Biology 13

Series Editors

Ph. Jeanteur, Y. Kuchino,
W.E.G. Müller (*Managing Editor*)
P.L. Paine

Ph. Jeanteur (Ed.)

Molecular and Cellular Enzymology

With 58 Figures

Springer-Verlag
Berlin Heidelberg New York
London Paris Tokyo
Hong Kong Barcelona
Budapest

Prof. Dr. PHILIPPE JEANTEUR
Institut de Génétique Moleculaire
C.N.R.S.-1919, Route de Mende
B.P. 5051
34033 Montpellier Cedex 01
France

ISBN-13:978-3-642-78583-2 e-ISBN-13:978-3-642-78581-8
DOI: 10.1007/978-3-642-78581-8

The Library of Congress has catalogued this serial publication as follows:
Library of Congress Catalog Card Number 75-79748

Softcover reprint of the hardcover 1st edition 1994

Typesetting: Macmillan India Ltd., Bangalore 25
39/3130/SPS – 5 4 3 2 1 0 – Printed on acid-free paper

Preface

For centuries, men have strived to unravel the mysteries of the living organism. They first described their surroundings, then the different parts of each living cell. In order to keep track of all their discoveries, they endeavoured to classify the observed objects (Linné 1700). Then, through these classifications, they proposed links between the numerous clusters they determined. With Lamarck and Darwin (1820–1860), the science of evolution was born. This part of biological science was purely descriptive and intuitive. In the minds of most scientists, living organisms could not be ruled by such simple physical or chemical laws (cf. Pasteur 1864).

However, chemists and later on, physicists became convinced that the same basic laws they had described for inert matter should also hold true for living organisms. The chemical synthesis of urea provided one of the very first clues to the universality of these laws. During the nineteenth and the beginning of the twentieth century, works by Berzelius (the concept of catalysis, 1850), Buchner (purification of the first enzyme, the zymase, 1904), Henry (1908), and Michaelis and Menten (the enzyme as a catalyst, 1912) were the milestones which paved the road to modern enzymology.

The first half of the twentieth century was marked by the elucidation of most metabolic pathways and by the emergence of protein biochemistry (essentially purification and characterization of proteins) and classical enzymology. The other components of the cells were regarded either as structural elements merely serving compartmentalization purposes (lipids) or as energy storage compounds (glucides), until DNA was equated with the genetic material in the mid-1940s (work by Avery, Delbrück, and others) and its double-helical structure established by Watson and Crick in 1953. Long before molecular biology reached its current level of power and sophistication, Jacob and Monod introduced the concept of regulation that was to keep biologists busy for (at least) the second half of the twentieth century. The cell was no longer considered a test tube where mother Nature was practicing chemistry, but as a small computer able to process and manage huge amounts of information. Therefore, questions in biology shifted from: "What are the components of the cell and which reactions do they catalyze?" to "What is the program of a given cell and what apparatus can carry out such a program?".

In a computer, microchips are composed of millions of transistors, each of which participates in only a limited number of elementary devices. However, the various functions that a microchip can perform depend on the design of the microchip. Therefore, a transistor is not important as such, but must be at the right place and act at the right time. In order to understand a living cell, the localization and the behaviour of each cell component in its local environment must be understood (among others, the proteins, the transistors of the cell, and the metabolites, the electrons of the cells).

To approach the above questions which open a new area of enzymology, new tools have to be developed. In Chapter I, J. Ricard shows that through mathematical modeling many dynamic processes which occur in vivo, in the living cell, or which may be observed in biomimetic systems such as bioreactors may be understood thanks to concepts directly derived from physical chemistry. In Chapter II, J. Haiech shows how genetic engineering, classical genetics and structural biology can be combined to allow the analysis of structure-function relationships in vivo. These new approaches also allow one to test evolutionary hypotheses and to directly determine where, on a given protein, the pressure of selection is important. Finally, in Chapter III, P. Viallet describes the latest progress in video-microspectrofluoroscopy. Such techniques have the potentiality to directly observe the localization and the dynamic behaviour of proteins and metabolites in a given cell. Such observations, initiated by the pioneering work of Kohen (1964), have pointed out that enzymatic studies performed in a test tube may be misleading when extrapolated to a cell. The local environment of an enzyme may dramatically change its behaviour and reveal totally unexpected properties which make them depart from the rules of classical enzymology.

Marseille and Montpellier, France
November 1993

J. Haiech
Ph. Jeanteur

Contents

List of Contributors

Addresses are given at the beginning of the respective contribution

Dynamics of Enzyme Reactions and Metabolic Networks in Living Cells. A Physico-Chemical Approach

Jacques Ricard[1], Guillermo Mulliert[1], Nicolas Kellershohn[1], Marie-Thérèse Giudici-Orticoni[1]

1. Introduction

The fine tuning of enzyme reactions is no doubt essential for the economy of the living cell. Today the deciphering of metabolic processes requires not only a quantitative knowledge about enzyme reactions that occur in the different compartments of the cell but also a quantitative understanding of how the dynamics of enzyme reactions and enzyme networks proceeds and is controlled *in vivo*. This quantitative understanding may only be achieved through the analysis of physical models of enzyme reactions and enzyme processes occurring in organized media. Perturbations of this complex dynamics must inevitably lead to alterations of metabolism and possibly to the death of the cell.

The dynamics of isolated enzyme reactions in dilute solutions is now well documented and the mechanisms of catalysis are known (Fersht et al. 1986; Jencks 1969; Jencks 1975; Koshland 1962; Koshland et al. 1972; Leatherbarrow et al. 1985; Lienhard et al. 1972; Lumry 1959; Storm and Koshland 1970; 1972 a and b; Secemski et al. 1972; Ho and Fersht 1986) One may wonder however whether these molecular mechanisms are sufficient *per se* to understand how an enzyme works within a living cell.

There are in fact different physical reasons which explain why an enzyme may

[1]Institut Jacques Monod, C.N.R.S. - Université Paris VII, 2 Place Jussieu - Tour 43, 75251 Paris CEDEX 5, France

not behave in the same way in dilute solution, in a test tube, and in a living cell. These reasons will be considered in the present monograph.

A first reason for this difference of behaviour is the existence of compartmentalization which may imply that the enzyme reaction be a vectorial process where catalysis is associated with the transport of matter through membranes. Disfavoured enzyme reactions may thus be driven far away from thermodynamic equilibrium. The synthesis of ATP by membrane-bound ATP-synthase typically belongs to this type of reaction.

A second physical reason which explains why an enzyme may behave in different ways, depending on wheter it is in free solution or in organized media, is the existence of a coupling that may exist between the diffusion of reactants and the enzyme reaction rate. In dilute, stirred, enzyme solutions this coupling obviously does not exist, for the diffusion is much faster than the enzyme reaction rate. In organized, unstirred viscous media, such as those occurring in living cells, the rate of diffusion is considerably slowed down, and therefore coupling may occur between diffusion and enzyme reaction rate. This implies that measuring, under these conditions, the rate of product formation does not yield the intrinsic properties of the enzyme but of the system resulting from the coupling between diffusion and reaction alone. From this coupling may emerge properties that are novel with respect to those of diffusion and of enzyme reaction. These coupled reactions may probably offer the simplest example of a molecular device able to sense not only the intensity of a signal but also the direction of change of that signal. In short this implies that this molecular device may be viewed as a simple biosensor.

A third reason for the difference of behaviour of membrane-bound enzymes with respect to their counterparts in free solution is to be found in the electrostatic interactions that may occur between the fixed charges of the membrane and the mobile charges. If the substrate of the bound enzyme for instance, is an ion, repulsion or attraction of this ion by the fixed charges of the membrane will generate apparent alterations of the kinetics of this bound enzyme. Moreover these alterations will be modulated by the ionic strength, which then becomes a modulator of the enzyme reaction. Another problem emerges from the study of electrostatic effects on bound-enzyme systems. This problem is to know whether the degree of spatial order which may, or may not, prevail in the distribution of enzyme molecules and fixed charges in the polyelectrolyte matrix, affects the overall response of the enzyme system. Enzymes that are bound to cell walls represent good examples that allow one to study experimentally these types of effects.

The interactions that have been considered above, and that may alter the dynamic behaviour of an enzyme reaction, are in fact physical interactions between enzyme molecules and their environment. But there may also exist purely functional interactions between enzyme reactions. In a metabolic network, where a chemical intermediate is released by an enzyme reaction and taken up as a substrate by the next one, the dynamic behaviour of the whole system may then be completely different from that of any of the individual enzyme reactions that are interconnected. Thus the network may possibly display sustained oscillations. These oscillations may either emerge from intrinsic properties of enzyme reactions that are interconnected, or from the interactions of these reactions with the cellular milieu.

The last reason which may explain the difference of behaviour existing between enzymes in free solution and in the living cell is the existence of multi-enzyme complexes in most eukaryotic cells. Many enzymes that catalyse consecutive reactions are in fact physically associated *in vivo*. Then one may expect the kinetic, or the dynamic, behaviour of these enzymes to be different from what one would have expected if these enzymes were physically distinct.

All these interactions, physical or functional, that may exist between enzymes and their environment may generate properties that are totally unexpected in classical enzymology. These properties may represent the physical basis for the storage and the transfer of information by living cells. They may be studied and understood most easily with the help of simple physical models. This monograph is specifically devoted to these new topics of cellular enzymology.

2. Flows and forces, diffusion, partition of mobile ions by charged matrices

Let there be a chemical reaction

$$\nu_1 X_1 + \ldots + \nu_r X_r + \ldots + \nu_k X_k \rightleftharpoons \nu_{k+1} X_{k+1} + \ldots + \nu_p X_p + \ldots + \nu_m X_m \tag{1}$$

where ν_r and ν_p are the stoichiometric coefficients of reactants and products, respectively. If ξ is the advancement of the reaction, n_r and n_p the number of moles of reactants and products, one has (see for instance Castellan 1971; Schnakenberg 1981)

$$\begin{aligned} -\, dn_r &= \nu_r \, d\xi \\ dn_p &= \nu_p \, d\xi \end{aligned} \tag{2}$$

If the reaction occurs at constant volume, thermodynamics allows one to write

$$dS = -\frac{1}{T} d\,G = -\frac{1}{T}\left(\sum_r \mu_r \, dn_r + \sum_p \mu_p \, dn_p\right) \tag{3}$$

where S, G, T and μ are the entropy, the Gibbs free energy, the absolute temperature and the chemical potential, respectively. Inserting expressions (2) into equation (3) yields

$$dS = -\frac{1}{T} dG = \frac{1}{T} d\xi \left(\sum_r \nu_r \mu_r - \sum_p \nu_p \mu_p \right) \tag{4}$$

The affinity of the reaction, A, is defined as

$$A = \sum_r \nu_r \mu_r - \sum_p \nu_p \mu_p = -\left(\frac{\partial G}{\partial \xi} \right)_{T,V} \tag{5}$$

and the rate of entropy production is then

$$\frac{dS}{dt} = \frac{1}{T} A \frac{\partial \xi}{\partial t} \tag{6}$$

Therefore when the affinity is null the rate of entropy production is null as well and the system is at equilibrium. Thus the affinity appears as a driving force which tends to pull the system towards equilibrium. The degree of advancement of the reaction, $d\xi/dt$, may itself be considered as a reaction flow, and it is therefore of interest to know which kind of simple explicit relation may occur between a flow and a force.

Let there be for instance the simple chemical reaction

$$A + B \underset{k'}{\overset{k}{\rightleftharpoons}} C \tag{7}$$

its velocity is

$$\frac{1}{V} \frac{d\xi}{dt} = k\, c_A\, c_B - k'\, c_C \tag{8}$$

where c_A, c_B and c_C are the concentrations of the reactants. Setting

$$Q = \frac{c_C}{c_A\, c_B} \qquad \text{and} \qquad K = \frac{k}{k'} \tag{9}$$

equation (8) assumes the form

$$\frac{1}{V} \frac{d\xi}{dt} = k\, c_A\, c_B \left(1 - \frac{Q}{K} \right) \tag{10}$$

As a chemical potential may be expressed as

$$\mu = \mu^{o} + RT \ln c \tag{11}$$

where μ^o is a standard chemical potential, the affinity of the reaction assumes the form

$$A = -\left\{\mu_C^o - \left(\mu_A^o + \mu_B^o\right) + RT \ln \frac{c_C}{c_A c_B}\right\} \tag{12}$$

The standard free energy difference, ΔG^o, is defined as

$$\Delta G^o = \mu_C^o - \left(\mu_A^o + \mu_B^o\right) \tag{13}$$

and this is equivalent to

$$\Delta G^o = -RT \ln K \tag{14}$$

Therefore the affinity of the reaction may be reexpressed as

$$A = -RT \ln \frac{Q}{K} \tag{15}$$

When $Q = K$, that is when the chemical reaction occurs under thermodynamic equilibrium, the affinity of the reaction is equal to zero, and from equation (10) it appears evident that the degree of advancement of the reaction, that is the reaction flow, is equal to zero as well. In the case of chemical reaction (7) the general explicit relation which associates a flow and a force is then

$$\frac{d\xi}{dt} = V k c_A c_B \left\{1 - \exp(-A/RT)\right\} \tag{16}$$

However, close to equilibrium, this exponential relation becomes a linear one, for

$$\exp(-A/RT) \approx 1 - \frac{A}{RT} \tag{17}$$

Then the flow-force relationship is

$$\frac{d\xi}{dt} = \frac{V k c_A c_B}{RT} A \tag{18}$$

This linear relation between flows and forces may be extended to diffusion processes. The first law of Fick predicts that the diffusion flow of molecules that pass across a surface should be proportional to the concentration gradient. One has

$$J = \frac{\partial n}{\partial t} = -DA_d \frac{\partial c}{\partial x} \tag{19}$$

where J is the diffusion flow, n the mole number, D the transport coefficient (in

$cm^2\ s^{-1}$), A_d the area of the surface through which diffusion occurs (in cm^2), c the concentration and x the distance (in cm).

The second law of Fick expresses the variation of the concentration of molecules that pass across a volume element. One has

$$\frac{\partial c}{\partial t} = D\frac{\partial^2 c}{\partial x^2} \tag{20}$$

If one assumes that diffusion is in steady state within this volume element

$$\frac{\partial^2 c}{\partial x^2} = 0 \tag{21}$$

which is equivalent to

$$\frac{\partial c}{\partial x} = -\frac{c_o - c_i}{l} \tag{22}$$

where c_o is the concentration of the solute on the cis side of the membrane, c_i its corresponding concentration on the trans side , and l the thickness of the membrane. Under steady state conditions, the diffusion flow, that is the number of moles that pass across a surface, is (first Fick's law)

$$J = \frac{DA_d}{l}\left(c_o - c_i\right) \tag{23}$$

The ratio DA_d/l is expressed in $cm^3\ s^{-1}$. It has thus the dimension of a volume, D_v, times a first order rate constant, h_d. Therefore

$$\frac{DA_d}{l} = D_v\, h_d \tag{24}$$

The diffusion flow per unit volume, J_v, is thus

$$J_v = \frac{J}{D_v} \tag{25}$$

Equation (23) may then be rewritten as

$$J_v = h_d\left(c_o - c_i\right) \tag{26}$$

and is thus expressed in "quasi-chemical" formalism. This is important for it appears that diffusion processes and chemical reactions may be represented in the same way and with the same units.

It is evident that the driving force of a diffusion process across a membrane is the difference of chemical potentials between the cis and the trans side of that membrane, $\mu_o - \mu_i$. In order to derive the flow-force relationship for such a diffusion process, one may rewrite equation (26) as

$$J_v = h_d \left\{ \exp\left(\frac{\mu_o - \mu^o}{RT}\right) - \exp\left(\frac{\mu_i - \mu^o}{RT}\right) \right\} \tag{27}$$

where μ^o is still the standard chemical potential. This equation may again be rewritten as

$$J_v = h_d \exp\left(-\frac{\mu^o}{RT}\right) \left\{ \exp\left(\frac{\mu_o}{RT}\right) - \exp\left(\frac{\mu_i}{RT}\right) \right\} \tag{28}$$

Setting

$$<\mu> = \frac{\mu_o + \mu_i}{2}$$

One has

$$J_v = h_d \exp\left(-\frac{\mu^o}{RT}\right) \exp\left(\frac{<\mu>}{RT}\right) \left\{ \exp\left(\frac{\mu_o - \mu_i}{2RT}\right) - \exp\left(-\frac{\mu_o - \mu_i}{2RT}\right) \right\} \tag{29}$$

which is a typical flow-force relationship. If this force is small, that is if the system is close to thermodynamic equilibrium, this equation becomes close to linearity. One has

$$\begin{aligned} \exp\left(\frac{\mu_o - \mu_i}{2RT}\right) &\approx 1 + \frac{\mu_o - \mu_i}{2RT} \\ \exp\left(-\frac{\mu_o - \mu_i}{2RT}\right) &\approx 1 - \frac{\mu_o - \mu_i}{2RT} \end{aligned} \tag{30}$$

If these expressions are inserted into the equation (29) one finds

$$J_v = \frac{h_d}{RT} \exp\left(\frac{<\mu> - \mu^o}{RT}\right) (\mu_o - \mu_i) \tag{31}$$

and the flow-force relationship is now linear. The general conclusion of this reasoning is that, whatever the nature of the dynamic process considered, there exists, close to thermodynamic equilibrium, a linear relationship between flows and forces.

This conclusion may be supported by an even more general formulation. Let there be a flow, J_i, generated by different forces, $X_1, .., X_j,..$, one has

$$J_i = f(X_1, ..., X_j, ...) \tag{32}$$

Close to equilibrium this function may be expanded in Mac Laurin series, namely

$$J_i = \bar{J}_i + \sum_j \frac{\partial J_i}{\partial X_j} X_j + ... \tag{33}$$

where $\bar{J}_i$ is the equilibrium flow which is indeed null. Therefore, close to equilibrium the flow may be considered as a linear combination of forces, namely

$$J_i = \sum_j L_{ij} X_j \tag{34}$$

where L_{ij} are the coupling coefficients and are defined as

$$L_{ij} = \frac{\partial J_i}{\partial X_j} \tag{35}$$

Relation (34) may be extended to any number of flows through the matrix relation

$$\begin{bmatrix} J_1 \\ J_2 \\ --- \\ J_n \end{bmatrix} = \begin{bmatrix} L_{11} & L_{12} & ... & L_{1n} \\ L_{21} & L_{22} & ... & L_{2n} \\ ----- & ----- & & ---- \\ L_{n1} & L_{n2} & ... & L_{nn} \end{bmatrix} \begin{bmatrix} X_1 \\ X_2 \\ --- \\ X_n \end{bmatrix} \tag{36}$$

The reciprocity relationship of Onsager implies that

$$L_{ij} = L_{ji} \qquad (i, j, = 1, 2, ...n) \tag{37}$$

The concepts of flow and force and the quasi-chemical formulation of diffusion processes are absolutely essential to understand how compartmentalization of enzymes within the cell, as well as the coupling between diffusion and enzyme reactions may generate novel properties of molecular systems in cell organelles. But there is also a physical effect which is essential in order to understand the behaviour of bound enzyme systems. This effect is the electrostatic partitioning of ions by insoluble polyelectrolytes. (Engasser and Horvath 1974 a, b, c, 1975, 1976; Horvath and Engasser 1974; Douzou and Maurel 1977 a and b; Maurel and Douzou 1976; Ricard et al. 1981; Goldstein et al. 1964)

Let there be an insoluble polyelectrolyte matrix. Owing to the electrostatic attraction or repulsion of mobile ions, their concentration inside the matrix will be different from the one prevailing in the bulk phase. The electrochemical potentials of an ion inside the matrix (subscript i) and outside that matrix (subscript o) are

$$\mu_i = \mu^o + RT \ln (\gamma_i c_i) + z F \psi_i$$
$$\mu_o = \mu^o + RT \ln (\gamma_o c_o) + z F \psi_o \qquad (38)$$

where γ_i and γ_o are the activity coefficients, z the valence of this ion multiplied by ± 1 depending on this ion is a cation or an anion , ψ_i and ψ_o the corresponding electrostatic potentials. At thermodynamic equilibrium

$$\mu_o - \mu_i = 0 = RT \ \ln \frac{\gamma_o c_o}{\gamma_i c_i} + z F \left(\psi_o - \psi_i\right) \qquad (39)$$

which is equivalent to

$$\frac{1}{z} \ln \frac{\gamma_i c_i}{\gamma_o c_o} = \frac{F \Delta \psi}{RT} \qquad (40)$$

with $\Delta\psi = \psi_o - \psi_i$. If a cation $A^{z'+}$ and an anion $B^{z'-}$ (with $z = \pm z'$) are present in these two phases one has

$$\frac{1}{z'} \ln \frac{{}^A\gamma_i \, {}^Ac_i}{{}^A\gamma_o \, {}^Ac_o} = \frac{1}{z'} \ln \frac{{}^B\gamma_o \, {}^Bc_o}{{}^B\gamma_i \, {}^Bc_i} = \frac{F \Delta \psi}{RT} \qquad (41)$$

where ${}^A\gamma$, ${}^B\gamma$, ${}^Ac$, ${}^Bc$ represent the activity coefficients and the concentrations of the two ions. One may define an electrostatic partition coefficient, Π, as

$$\Pi = \exp\left(F\Delta\psi/RT\right) \qquad (42)$$

and expression (41) may be rewritten as

$$\left(\frac{{}^A\gamma_i \, {}^Ac_i}{{}^A\gamma_o \, {}^Ac_o}\right)^{1/z'} = \left(\frac{{}^B\gamma_o \, {}^Bc_o}{{}^B\gamma_i \, {}^Bc_i}\right)^{1/z'} = \Pi \qquad (43)$$

Let there be the two phases inside and outside a polyelectrolyte matrix containing different anions of different valence, B^-, B^{2-}, .., $B^{z'-}$ and different cations A^+, one has

$$\Pi = \frac{\Sigma B_o^-}{\Sigma B_i^-} = \ldots = \left(\frac{\Sigma B_o^{z'-}}{\Sigma B_i^{z'-}}\right)^{1/z'} = \frac{\Sigma A_i^+}{\Sigma A_o^+} \qquad (44)$$

Moreover electroneutrality should hold inside and outside the matrix. Therefore one has

$$\Sigma B_o^- + \ldots + z' \, \Sigma B_o^{z'-} = \Sigma A_o^+$$
$$\Sigma B_i^- + \ldots + z' \, \Sigma B_i^{z'-} \pm \Delta^{\pm} = \Sigma A_i^+ \qquad (45)$$

where $\Delta^{\pm}$ represents the fixed positive or negative charge density in the matrix depending on the polyelectrolyte is a polycation or a polyanion. The electroneutrality equation that applies to the matrix phase (the so-called local electroneutrality equation) may be rewritten as

$$\frac{\sum B_o^-}{\Pi} + \dots + \frac{z' \sum B_o^{z'-}}{\Pi^{z'}} \pm \Delta^{\pm} = \Pi \sum A_o^+ \tag{46}$$

and taking account of the first expression (45), equation (46) may be rewritten as

$$\Pi^{z'+1} \pm \frac{\Delta^{\pm}}{\sum B_o^- + \dots + z' \sum B_o^{z'-}} \Pi^{z'} - \dots - \frac{z' \sum B_o^{z'-}}{\sum B_o^- + \dots + z' \sum B_o^{z'-}} = 0 \tag{47}$$

Solving this equation allows one to obtain the expression of the partition coefficient as a function of anion concentrations and of fixed charge density. For the very simple situation, for instance, of only two ionic species A^+ and B^-, equation (47) above reduces to a quadratic, namely

$$\Pi^2 - \frac{\Delta^-}{B_o} \Pi - 1 = 0 \tag{48}$$

for a polyanion , and

$$\Pi^2 + \frac{\Delta^+}{B_o} \Pi - 1 = 0 \tag{49}$$

for a polycation. The expression Π is then, for a polyanion

$$\Pi = \frac{\Delta^-}{2B_o} + \frac{1}{2} \sqrt{\left(\frac{\Delta^-}{B_o}\right)^2 + 4} \tag{50}$$

and for a polycation

$$\Pi = \frac{1}{2} \sqrt{\left(\frac{\Delta^+}{B_o}\right)^2 + 4} - \frac{\Delta^+}{2B_o} \tag{51}$$

In Figure 1 is shown the variation of the electrostatic partition coefficient of a polyanion as a function of B_o, for different values of Δ^-. For very high values of B_o, Π should approach unity.

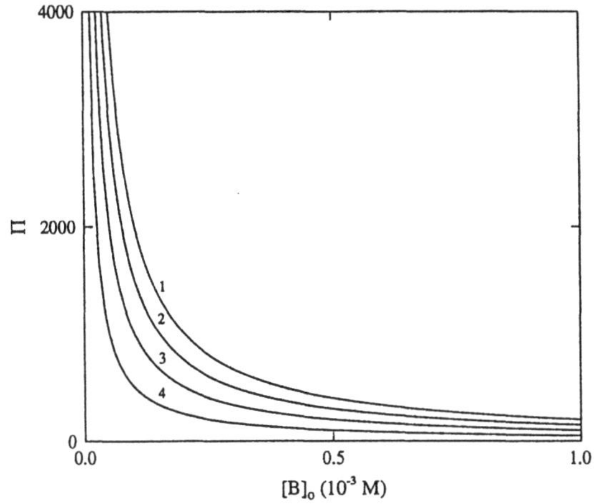

Fig. 1. Variation of the electrostatic partition coefficient, Π, as a function of the bulk anion concentration. Curves *1 - 4* pertain to Δ^- values equal to 0.2, 0.15, 0.1 and 0.05 respectively

3. Compartmentalization of enzyme reactions and the energy metabolism of the cell

A number of reactions that are strongly endergonic (negative affinity) if considered isolated, still proceed to completion within the living cell. These reactions do not violate thermodynamics and their occurrence may be explained by compartmentalization in the cell of enzyme molecules that are in charge of these kinetic processes. Probably the most striking example of the role played by compartmentalization in the occurrence of thermodynamically disfavoured reactions is the synthesis of ATP. This reaction is accompanied by a free energy change of about 5 kJ per mole and is conditioned by an enzyme called ATP-synthase. Owing to the strongly endergonic character of this reaction, isolated ATP-synthase can only catalyse the hydrolysis of ATP, but not its synthesis. This enzyme however is not isolated in the cell but is bound to the internal membrane of mitochondria, of bacteria and to the membrane of thylakoids. This synthesis of ATP is in fact coupled to a proton flow across the membrane and it is therefore important to understand how this coupled process may occur without violating thermodynamics and still results in the occurrence of a strongly disfavoured chemical reaction.

Let us consider a permeable membrane that separates two compartments called cis (') and trans ("), both with the same volume, V, and containing different kinds of molecules with chemical potentials μ_i. If the temperature and the potentials are different in the two compartments a flow of energy and several flows of matter will occur. As the two compartments have the same volume (Figure 2), one has (see for instance Hill 1977; Hill and Chen 1970)

$$\begin{aligned} dE' &= T'dS' + \sum_i \mu_i' dn_i' \\ dE'' &= T''dS'' + \sum_i \mu_i'' dn_i'' \end{aligned} \tag{52}$$

Moreover since the entropy is an extensive function one has

$$dS = dS' + dS'' \tag{53}$$

The principles of energy and mass conservation imply that

$$\begin{aligned} dE' &= -dE'' \\ dn_i' &= -dn_i'' \end{aligned} \tag{54}$$

One may thus write

$$dS = dS' + dS'' = dE''\left(\frac{1}{T''} - \frac{1}{T'}\right) + \sum_i \left(\frac{\mu_i'}{T'} - \frac{\mu_i''}{T''}\right) dn_i'' \tag{55}$$

The forces which drive the flows of energy and matter from one compartment to the other are thus

$$\begin{aligned} X_E &= \frac{1}{T''} - \frac{1}{T'} \\ X_i &= \frac{\mu_i'}{T'} - \frac{\mu_i''}{T''} \end{aligned} \tag{56}$$

These forces may indeed be positive or negative. The flows of energy and matter are

$$\begin{aligned} J_E &= \frac{dE''}{dt} \\ J_i &= \frac{dn_i''}{dt} \end{aligned} \tag{57}$$

It then follows from expressions (55), (56) and (57) that the dissipation of entropy during the transport of energy and matter is

$$\frac{dS}{dt} = J_E X_E + \sum_i J_i X_i \geq 0 \tag{58}$$

This relation is important for it shows that molecules may be transported against a concentration gradient provided a flow of matter is coupled with a flow of energy. Alternatively a chemical reaction of negative affinity (an endergonic reaction) may be driven to completion if it is coupled with a transport of matter.

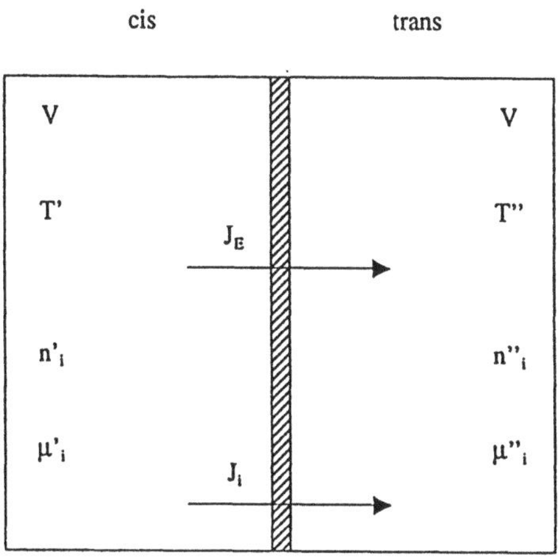

Fig. 2. Transport of matter and energy across a membrane (see text).

This situation, which has been discussed from a thermodynamic viewpoint , is precisely the one which occurs in the synthesis of ATP by ATP-synthase. In mitochondria electron transfer from NADH to molecular oxygen results in the transfer of four protons in the intermembrane space (Figure 3).

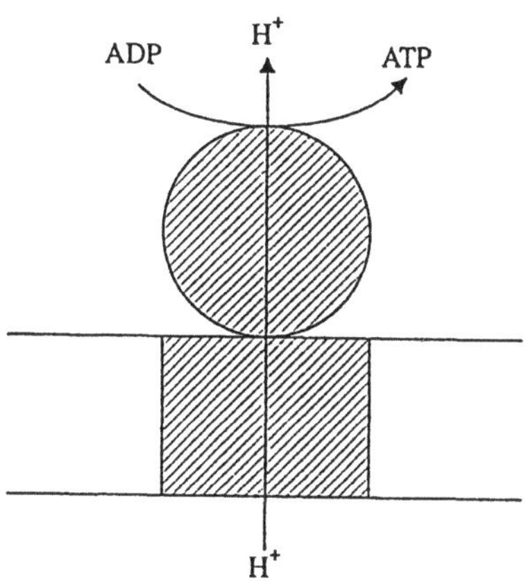

Fig. 3. Schematic picture of ATP-synthase of mitochondria

This accumulation of protons is an active process for the protons are transferred against a concentration gradient. Proton concentration in the intermembrane space then becomes higher than in the matrix. ATP-synthase is a multi-subunit, high molecular weight protein borne by the internal mitochondrial membrane. This enzyme has a globular head bearing the active site and a stalk buried in the membrane. The head is located outside the membrane, in direct contact with the matrix. Both the stalk and the globular head are traversed by a pore. As the protons accumulate in the inter-membrane space, they diffuse through the pore towards the matrix (Figure 3) and this diffusion process is coupled to ATP synthesis . As the affinity (or the force) associated to ATP synthesis is strongly negative the chemical reaction leading to the synthesis of ATP cannot proceed to any significant extent if considered in isolation. Equation (58) above shows this reaction becomes possible if coupled to proton transfer.

The situation is basically similar for the synthesis of ATP in bacterial cells and in chloroplast thylakoids. In bacteria the globular head of ATP-synthase is directed towards the interior of the cell. Protons are expelled outside the cell in the course of the electron transfer process. Therefore their concentration outside the cell becomes higher than inside that cell. Then the protons tend to diffuse through the pore of ATP-synthase towards the interior of the bacterial cell (Figure 4). As previously this diffusion is coupled with the synthesis of ATP.

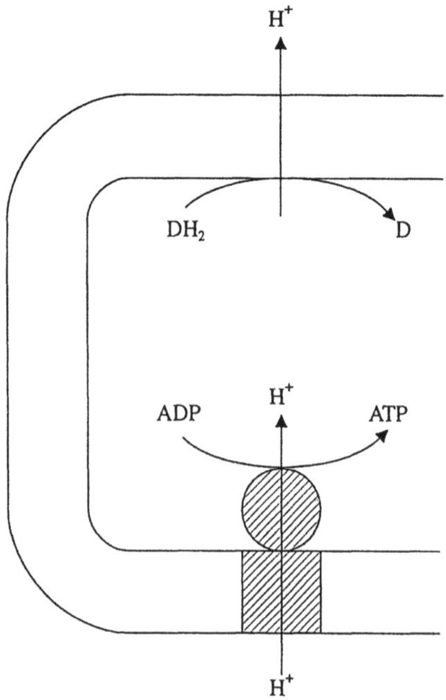

Fig. 4. Schematic picture of ATP-synthase in bacteria

The thylakoids are vesicles stacked in chloroplasts and which contain chlorophyll. The globular head of ATP-synthase is directed towards the outside of the vesicles. The light excitation of chlorophyll and the resulting electron transfer generate proton accumulation inside the thylakoid vesicles where the pH declines. Again protons tend to diffuse through the pore of ATP-synthase and this is coupled to the synthesis of ATP (Figure 5).

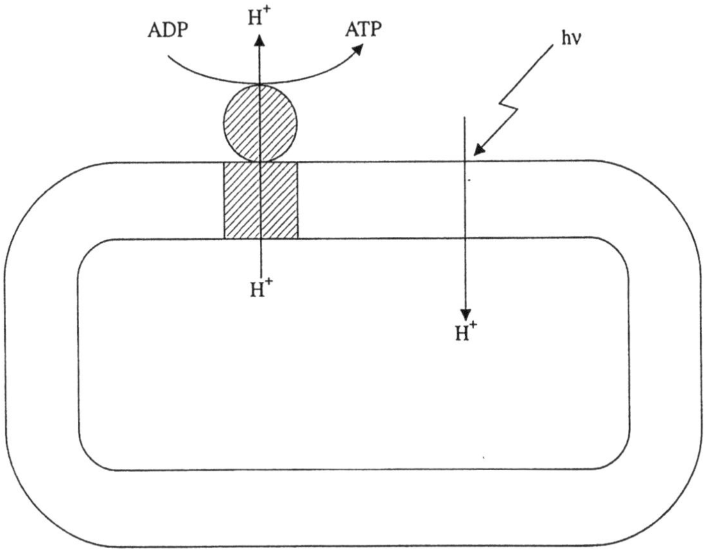

Fig. 5. Schematic picture of ATP-synthase in thylakoids

Whatever the biological system considered, the synthesis of ATP is driven by a proton-motive force, Δp. This proton-motive force is defined from the expression of the electrochemical potential, μ_H

$$\mu_H = \mu^o + RT \ln[H] + F\psi \tag{59}$$

The Gibbs free energy associated with the transfer of a proton through a pore of the ATP-synthase is

$$\mu_o - \mu_i = \Delta\mu_H = -2.3 \; RT\left(pH_o - pH_i\right) + F\left(\psi_o - \psi_i\right) \tag{60}$$

and the proton-motive force is then

$$\Delta_p = \frac{\Delta\mu_H}{F} = -2.3 \; \frac{RT}{F} \left(pH_o - pH_i\right) + \left(\Psi_o - \Psi_i\right) \tag{61}$$

It is thus the difference of pH and of electrostatic potential that drives the chemical reaction. It is therefore the compartmentalization of ATP-synthase which is responsible for the synthesis of ATP (Mitchell 1966, 1968, 1979; Mitchell et al. 1979, Kell and Westerhoff 1984).

4. Coupling between reactant diffusion and bound enzyme reaction rate

Within a living cell the viscosity may be high enough to result in a decrease of the diffusion rate of reactants. This decrease is called diffusional resistance. If one of these reactants is involved in an enzyme reaction, the corresponding reaction rate may be lower than the diffusion rate. Under these conditions, there exists a coupling between diffusion and enzyme reaction. From this coupling, novel kinetic properties may be expected to occur (Engasser and Horvath 1974 a,b and c; Hervagault and Thomas 1985; Hervagault et al. 1984; Ricard and Noat 1984 a and b; Harrison 1987, Thomas et al. 1977).

The very existence of coupling between diffusion and enzyme reaction is based on the possibility of representing a diffusion process with the quasi-chemical formalism. If one considers for instance enzyme molecules bound to an impermeable surface, there may exist in the vicinity of this surface a gradient of substrate concentration. If the overall system is in steady state, the diffusion flow assumes the form

$$J_V = h_d \left\{ S(o) - S(m) \right\} \tag{62}$$

and if the enzyme follows Michaelis-Menten kinetics, the corresponding enzyme reaction rate is

$$v_e = \frac{V\,S\,(m)\,/\,K}{1 + S\,(m)\,/K} \tag{63}$$

In these expressions V and K are the Vmax and the Km of the enzyme reaction, S(o) and S(m) the substrate concentration at the distance o and at the distance m, respectively. The origin of the space co-ordinate, o, is located at the frontier of the bulk phase, where the substrate concentration does not vary in space. From a thermodynamic viewpoint, this region may be considered as a reservoir. The distance m thus corresponds to the distance that separates the reservoir from the surface where the enzyme molecules are located. The substrate gradient is linear over the distance $x \in [o, m]$ and this is a consequence of the steady state which is assumed to occur.

Coupling between diffusion and enzyme reaction means that

$$h_d \left\{ S(o) - S(m) \right\} = \frac{V\,S(m)/K}{1 + S(m)/K} \tag{64}$$

There is an obvious advantage in writing this equation in dimensionless form by setting

$$s_\sigma = \frac{S(m)}{K} \qquad s_o = \frac{S(o)}{K} \qquad h_d^* = \frac{h_d K}{V} \tag{65}$$

and the coupling equation assumes the simpler form

$$h_d^* (s_o - s_\sigma) - \frac{s_\sigma}{1 + s_\sigma} = 0 \tag{66}$$

which is equivalent to

$$h_d^* s_\sigma^2 + \left\{ h_d^* \left(1 - s_o\right) + 1 \right\} s_\sigma - h_d^* s_o = 0 \tag{67}$$

The positive root of this equation is

$$s_\sigma = \frac{\sqrt{\left\{1 + h_d^* \left(1 - s_o\right)\right\}^2 + 4 h_d^{*2} s_o} - \left\{h_d^* \left(1 - s_o\right) + 1\right\}}{2 h_d^*} \tag{68}$$

and expresses how the local dimensionless substrate concentration, s_σ, varies with respect to the bulk dimensionless substrate concentration, s_o. This concentration s_o is the only one which may be known directly, whereas s_σ is the one "seen" by the enzyme molecules. It is obvious from equation (68) that if the enzyme *per se* follows Michaelis-Menten kinetics with respect to s_σ, which is the hypothesis which has been considered above, it cannot follow the same type of kinetics with respect to the bulk concentration, s_o, which is the only variable that may be known by experiment. In fact this kinetics resembles positive co-operativity. At "low" substrate concentration, the overall coupled system is limited by the diffusion rate, whereas at "high" substrate concentration it is limited by the enzyme activity.

A far more interesting situation from a biological viewpoint is to be expected if the enzyme reaction rate displays terms that are non-linear in substrate concentration. The simplest situation of this type is the one encountered with an enzyme that is inhibited by an excess of substrate. This enzyme equation assumes the form

$$v = \frac{VS(m)/K}{1 + S(m)/K + K_2 S^2(m)/K} \tag{69}$$

where K_2 is the affinity constant of the substrate for the enzyme-substrate complex. Setting $\lambda = K_2K$, the dimensionless coupling equation becomes

$$h_d^* (s_o - s_\sigma) - \frac{s_\sigma}{1 + s_\sigma + \lambda\, s_\sigma^2} = 0 \tag{70}$$

which may be rewritten as

$$h_d^* \lambda\, s_\sigma^3 - h_d^* (\lambda\, s_o - 1)\, s_\sigma^2 + \{h_d^* (1 - s_o) + 1\}\, s_\sigma - h_d^*\, s_o = 0 \tag{71}$$

This equation may have three real positive roots (Descartes rule of signs). Therefore upon plotting s_σ as a function of s_o one may obtain the type of curve shown in Figure 6.

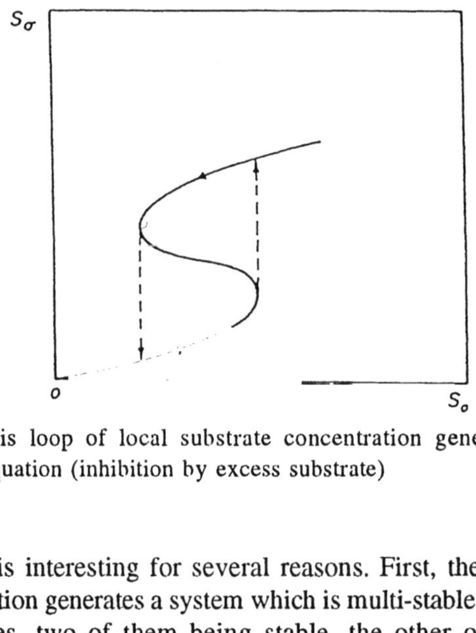

Fig. 6. A hysteresis loop of local substrate concentration generated by non-linear terms of the rate equation (inhibition by excess substrate)

This situation is interesting for several reasons. First, the coupling between diffusion and reaction generates a system which is multi-stable and displays in fact three steady states, two of them being stable, the other one being unstable. Second, the system is able to sense not only the value of a concentration in the bulk phase but also whether this concentration is reached after increasing or decreasing a former concentration. This implies that the system is able to "recall" the values of the concentrations that were anterior to that occurring at time t. That is, the system is able to store short-term memory and therefore displays hysteresis effects. Third, these effects are specifically due to the occurrence of at least one non-linear term in the rate equation. Last but not least all these effects must disappear under equilibrium conditions. This model probably represents the simplest one that allows sensing of chemical signals and storing information by enzymes.

5. Electric partitioning of ions and reaction rate of bound enzyme systems

Electric partitioning of mobile ions by the fixed charges of a polyelectrolyte matrix may generate interesting kinetic effects if the enzyme molecules are buried in this matrix. Probably the more interesting of these effects is observed if the substrate itself is an ion. Then, one may expect electrostatic attraction or repulsion of substrate to generate alterations of the kinetics of the bound enzyme. These alterations have been termed electrostatic co-operativity. As biomembranes and cell envelopes are polyelectrolyte matrices, one may expect these effects to be important *in vivo*. This matter will be discussed in detail in the next section.

5.1. Electrostatic co-operativity of a bound enzyme system

If molecules of an enzyme that follows Michaelis-Menten kinetics are homogeneously distributed in a polyelectrolyte matrix, the reaction rate that may be measured experimentally is

$$v = \frac{V_m S_i}{K_m + S_i} \tag{72}$$

where S_i is still the local substrate concentration, inside the matrix. Indeed this equation implies that no diffusional resistance of the substrate occurs in this heterogeneous medium and therefore that the substrate diffuses rapidly inside and outside the matrix. As the local concentration of the substrate, S_i, is unknown, one may express the reaction rate as a function of S_o. Taking advantage of the definition of the electrostatic potential Π, one has

$$v = \frac{V_m S_o}{K_m \Pi^{z'} + S_o} \tag{73}$$

where S_o is now the bulk substrate concentration. When this substrate concentration S_o is raised, Π varies as well in such a way that equation (73) is not the equation of a rectangular hyperbola. In other words, the bound enzyme no longer follows Michaelis-Menten kinetics. The departure from this type of kinetics mimics positive co-operativity if there is electrostatic repulsion of the substrate by the fixed charges of the matrix, or negative co-operativity if there is electrostatic attraction of the substrate inside the matrix. A particularly interesting case occurs if the matrix is a polyanion and if the substrate is a monoanion. Then the expression of the electrostatic partition coefficient is equation (50) which, for strong repulsion effects, may be approximated to (Ricard et al. 1981)

$$\Pi = \frac{\Delta^{\cdot}}{S_o} \tag{74}$$

Then the reaction rate equation assumes the form

$$\frac{V_m S_o^2}{K_m \Delta^{\cdot} + S_o^2} \tag{75}$$

with respect to S_o the bound enzyme thus displays sigmoidal kinetics. For various reasons that will appear evident in the next section, it is advantageous to define a new variable σ_o as

$$\sigma_o = \frac{S_o^2}{K_m} \tag{76}$$

which has the dimension of a concentration. With respect to σ_o the reaction rate follows a rectangular hyperbola.

Another important consequence of this reasoning is that ionic strength which may have no effect on the enzyme reaction when the enzyme is in free solution, must behave as an activator, or as an inhibitor if the enzyme is bound to a polyelectrolyte. As shown in Figure 1, raising the ionic strength results in a variation of the electrostatic partition coefficient Π which tends to unity. Then the bound enzyme behaves exactly as if it were free in solution.

5.2. Spatial organization of fixed charges and enzyme molecules as a source of co-operativity

In the previous section it has been considered that the enzyme molecules and the fixed charges are randomly, or homogeneously, distributed in the matrix. This may well not be the case and one may raise a novel and important question, namely to know whether the overall enzyme reaction rate that can be measured depends upon the type and the degree of spatial organization of the fixed charges and of the enzyme molecules (Ricard et al. 1989).

Therefore, at this stage, it is important to know whether it is possible to express quantitatively the degree of spatial order that may , or may not, exist in the distribution of fixed charges and enzyme molecules. It is also important to realize which significance is given to the term organization. In the following, this term is taken to mean the lack of pure randomness, or the presence of macroscopic heterogeneity, in the spatial distribution of fixed charges and enzyme molecules. If the fixed charges and the enzyme molecules are clustered in the matrix, this represents some form of spatial organization. One may express quantitatively the degree of organization by the monovariate moments of charge and enzyme distribution as well as by the bivariate moments that associate these two distributions.

One may express the charge density, Δ_i, of the charge clusters with respect to its mean, $< \Delta >$, and alternatively the maximum reaction velocity, V_j, (proportional to enzyme density) relative to its mean, $< V >$. One has thus

$$\begin{aligned} \Delta_i &= < \Delta > + \delta_i \qquad (i = 1, ..., n) \\ V_j &= < V > + \varepsilon_j \qquad (j = 1, ..., n) \end{aligned} \tag{77}$$

where δ_i and ε_j are the deviations relative to their corresponding means. If they are normally distributed about the zero value the moments of odd degree are null and one has thus

$$\begin{aligned} \sum_i \sum_j f_{ij}\delta_i &= \sum_i f_i\delta_i = N\mu_1(\delta) = 0 \\ \sum_i \sum_j f_{ij}\delta_i^2 &= \sum_i f_i\delta_i^2 = N\mu_2(\delta) = N\,\mathrm{var}(\delta) \\ \sum_i \sum_j f_{ij}\delta_i^3 &= \sum_i f_i\delta_i^3 = N\mu_3(\delta) = 0 \\ \sum_i \sum_j f_{ij}\varepsilon_j &= \sum_j f_j\varepsilon_j = N\mu_1(\varepsilon) = 0 \end{aligned} \tag{78}$$

In these expression $\mu(\delta)$ and $\mu(\varepsilon)$ represent the monovariate moments of charge and enzyme density distributions and N the number of clusters.

The bivariate moments associate the charge and enzyme density distributions. They express whether the charge density is correlated with the enzyme density . They therefore represent a quantitative estimation of the degree of spatial organization of charges and enzyme molecules in the matrix . If the density of these scaled charges and enzyme molecules is normally distributed, some of these bivariate moments are null. The expression of these moments is

$$\begin{aligned} \sum_i \sum_j f_{ij}\delta_i\varepsilon_j &= N\mu_{1,1}(\delta,\varepsilon) = N\,\mathrm{cov}(\delta,\varepsilon) \\ \sum_i \sum_j f_{ij}\delta_i^2\varepsilon_j &= N\mu_{2,1}(\delta,\varepsilon) = 0 \\ \sum_i \sum_j f_{ij}\delta_i^3\varepsilon_j &= N\mu_{3,1}(\delta,\varepsilon) \end{aligned} \tag{79}$$

If the charges and enzyme molecules are clustered in the matrix the moments $\mu(\delta)$ and $\mu(\varepsilon)$ are both different from zero, except if the clusters have the same charge and the same enzyme density, which is a rather unlikely situation. If there is some sort of spatial organization of the charges with respect to the enzyme molecules, the bivariate moments $\mu(\delta,\varepsilon)$ will be different from zero.

The simplest type of spatial organization is the one which occurs when the charges and (or) the enzyme molecules are clustered in the matrix. From that respect, five types of spatial organization may be considered (Figure 7). In the first type (A) there is no organization. The charges and the enzyme molecules are randomly distributed in the matrix . Thus, there is no macroscopic heterogeneity in their spatial distribution . Therefore, in any macroscopic region of space the monovariate moments $\mu(\delta)$ and $\mu(\varepsilon)$ will be equal to zero as well as the bivariate moments $\mu(\delta,\varepsilon)$. In the second type of organization (B) it is assumed that the charges are clustered, but that the enzyme molecules are not. Hence, in general, $\mu(\delta) \neq 0$, but $\mu(\varepsilon) = 0$ and $\mu(\delta,\varepsilon) = 0$. The converse may indeed occur. That is the enzyme molecules are clustered, but the charges are not (C). Then $\mu(\delta) = 0$, $\mu(\delta,\varepsilon) = 0$, but in general $\mu(\varepsilon) \neq 0$. A higher type of organizational complexity occurs if both the fixed charges and the enzyme molecules are clustered, and clusters of charges partly overlap clusters of enzyme molecules (D). Then in all generality $\mu(\delta) \neq 0$, $\mu(\varepsilon) \neq 0$ and $\mu(\delta,\varepsilon) \neq 0$. The last type of organizational complexity occurs if the charges and the enzyme molecules are clustered and these two types of clusters exactly overlap (E). Then, again in all generality, the monovariate and bivariate moments are different from zero. These different types of spatial organization are depicted in Figure 7.

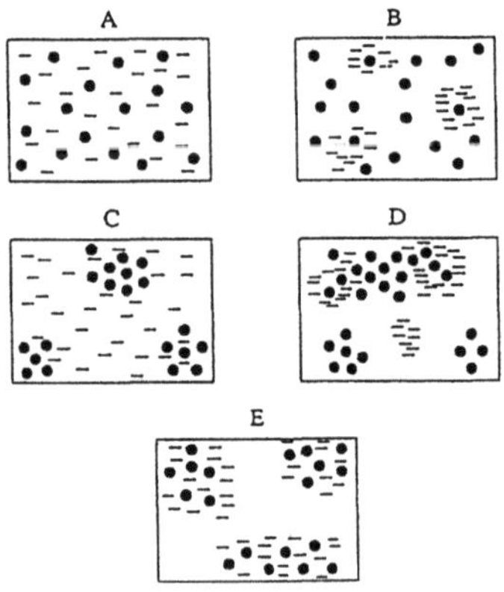

Fig. 7. Models of organization of fixed charges and enzyme molecules in a charged matrix. **A** The enzyme molecules and the fixed charges are randomly distributed in the matrix. **B** The enzyme molecules are randomly distributed but the fixed charges are clustered. **C** The enzyme molecules are clustered but the fixed charges are randomly distributed. **D** The enzyme molecules and the fixed charges are clustered, and the clusters partly overlap. **E** The enzyme molecules and the fixed charges are clustered and the clusters are superimposed

In order to express how these monovariate and bivariate moments affect the enzyme reaction rate, one has to consider first the situation where the clusters of charges and the clusters of enzyme molecules are exactly superimposed. Then for a monovalent substrate one has

$$v = \sum_i \sum_j \frac{f_{ij} V_j S_o}{K \Pi_i + S_o} \tag{80}$$

If the substrate is a monoanion, in such a way that expression (74) may be taken as a reasonable approximation of reality, equation (80) above may be written as

$$v = \sum_i \sum_j \frac{f_{ij} V_j S_o^2}{K \Delta_i + S_o^2} \tag{81}$$

If the new variable σ_o (see equation 76) is inserted into expression (81) one has

$$v = \sum_i \sum_j \frac{f_{ij} V_j \sigma_o}{\Delta_i + \sigma_o} \tag{82}$$

Under this form, it becomes evident that any departure from Michaelis-Menten kinetics with respect to σ_o is due to the spatial organization of the fixed charges, but not to that of the enzyme molecules. This new type of co-operativity can therefore be called organizational co-operativity.

Let v_{ij} be the reaction rate pertaining to a particular cluster, one has

$$v_{ij} = \left(< V > + \varepsilon_j\right) \frac{\sigma_o}{< \Delta > + \delta_i + \sigma_o} \tag{83}$$

Expanding this expression in Taylor series with respect to the variable δ_i yields

$$v_{ij} = \left(<V> + \varepsilon_j\right) \left\{ \frac{\sigma_o}{<\Delta> + \sigma_o} - \frac{\delta_i \sigma_o}{\left(<\Delta> + \sigma_o\right)^2} + \frac{\delta_i^2 \sigma_o}{\left(<\Delta> + \sigma_o\right)^3} - \ldots \right\} \tag{84}$$

One may define dimensionless variables as

$$\sigma_o^* = \frac{\sigma_o}{<\Delta>} \qquad \delta_i^* = \frac{\delta_i}{<\Delta>} \qquad \varepsilon_j^* = \frac{\varepsilon_j}{<V>} \tag{85}$$

and therefore equation (84) may be rewritten as

$$v_{ij} = \langle V \rangle \left(1 + \varepsilon_j^*\right) \left\{ \frac{\sigma_o^*}{1 + \sigma_o^*} - \frac{\delta_i^* \sigma_o^*}{\left(1 + \sigma_o^*\right)^2} + \frac{\delta_i^{*2} \sigma_o^*}{\left(1 + \sigma_o^*\right)^3} - \ldots \right\} \quad (86)$$

If, as previously assumed, δ_i and ε_j , are normally distributed, δ_i^* and ε_j^* are normally distributed as well, and their distribution is characterized by the monovariate and bivariate moments $\mu(\delta^*)$, $\mu(\varepsilon^*)$ and $\mu(\delta^*,\varepsilon^*)$.

The overall reaction rate may be obtained after summing up the rates v_i pertaining to the N clusters. One has

$$v = \sum_i \sum_j \langle V \rangle f_{ij} \left(1 + \varepsilon_j^*\right) \left\{ \frac{\sigma_o^*}{1 + \sigma_o^*} - \frac{\delta_i^* \sigma_o^*}{\left(1 + \sigma_o^*\right)^2} + \frac{\delta_i^{*2} \sigma_o^*}{\left(1 + \sigma_o^*\right)^3} - \ldots \right\} \quad (87)$$

It may be shown that δ_i^* is of necessity smaller than unity. Therefore the series in brackets is of necessity convergent. Developing equation (86) yields

$$\frac{v}{N \langle V \rangle} = \frac{\sigma_o^*}{1 + \sigma_o^*} \left\{ 1 - \frac{\mu_{1,1}\left(\delta^*, \varepsilon^*\right)}{1 + \sigma_o^*} + \frac{\mu_2\left(\delta^*\right) + \mu_{2,1}\left(\delta^*, \varepsilon^*\right)}{\left(1 + \sigma_o^*\right)^2} - \ldots \right\} \quad (88)$$

which again may be written in more compact form as

$$\frac{v}{N \langle V \rangle} = \lim_{m \to \infty} \frac{\sigma_o^*}{1 + \sigma_o^*} \left\{ 1 + \sum_{r=1}^{m} (-1)^r \frac{\mu_r\left(\delta^*\right) + \mu_{r,1}\left(\delta^*, \varepsilon^*\right)}{\left(1 + \sigma_o^*\right)^r} \right\} \quad (89)$$

Therefore the important consequence of these results is that the reaction rate not only depends upon the mean enzyme and charge density but also upon the spatial organization of the fixed charges and of the enzyme molecules.

If, for instance, the Taylor series converges rapidly enough to generate terms whose contribution is negligible above m = 2, the corresponding reaction rate reduces to

$$\frac{v}{N \langle V \rangle} = \frac{\sigma_o^*}{1 + \sigma_o^*} \left\{ 1 - \frac{\operatorname{cov}\left(\delta^*, \varepsilon^*\right)}{1 + \sigma_o^*} + \frac{\operatorname{var}\left(\delta^*\right)}{\left(1 + \sigma_o^*\right)^2} \right\} \quad (90)$$

It is thus evident that if the moments $\operatorname{var}(\delta^*)$ and $\operatorname{cov}(\delta^*,\varepsilon^*)$ are both equal to zero this equation reduces to a Michaelis-Menten process in σ_o^*. Any deviation with respect to this type of behaviour is thus the consequence of spatial organization of the fixed charges in the matrix. Equation (90) may be rewritten as

$$\frac{v}{N \langle V \rangle} = \frac{\sigma_o^*}{1+\sigma_o^*} + \Xi\left(\sigma_o^*\right) \tag{91}$$

where the function $\Xi\left(\sigma_o^*\right)$ is

$$\Xi\left(\sigma_o^*\right) = -\frac{\sigma_o^*}{\left(1+\sigma_o^*\right)^2}\operatorname{cov}\left(\delta^*, \varepsilon^*\right) + \frac{\sigma_o^*}{\left(1+\sigma_o^*\right)^3}\operatorname{var}\left(\delta^*\right) \tag{92}$$

Therefore this function expresses how the spatial organization of fixed charges modulates the enzyme reaction velocity. This function is called the function of organizational rate modulation. Equation (92) shows that this function may take only positive values if $\operatorname{cov}\left(\delta^*, \varepsilon^*\right) = 0$. This means that if there is no correlated spatial organization of fixed charges and enzyme molecules, the effect of spatial organization of charges alone is to enhance the reaction rate. Alternatively, if this correlated spatial organization occurs, the function $\Xi\left(\sigma_o^*\right)$ may take negative values and therefore correlated spatial organization of charges and enzyme molecules may generate a decrease of the enzyme reaction velocity. In either case, when σ_o^* increases, the effect of spatial organization on the enzyme reaction rate becomes negligible. These conclusions are pictured in Figure 8.

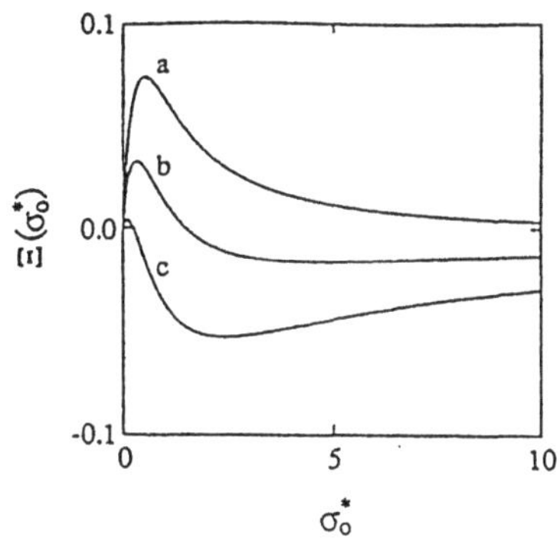

Fig. 8. Enhancement and inhibition of enzyme activity through spatial organization of fixed charges and enzyme molecules. When the function $\Xi(\sigma_o^*)$ takes positive values spatial arrangement of fixed charges results in an enhancement of the reaction rate an when this function adopts negative values the spatial organization of charges results in a decrease of the reaction rate. The $\Xi(\sigma_o^*)$ function is plotted *versus* the dimensionless variable σ_o^* for different values of $\operatorname{cov}(\delta^*,\varepsilon^*)$ and a fixed value of $\operatorname{var}(\delta^*) = 0.6$. Curve *a*: $\operatorname{cov}(\delta^*,\varepsilon^*) = 0$, curve *b*: $\operatorname{cov}(\delta^*,\varepsilon^*) = 0.3$, curve *c*: $\operatorname{cov}(\delta^*,\varepsilon^*) = 0.6$

The co-operativity generated by the spatial organization of fixed charges and enzyme molecules may be evaluated through the Hill function $h(\sigma_o^*)$

$$h(\sigma_o^*) = 1 + \Omega(\sigma_o^*) \tag{93}$$

where $\Omega(\sigma_o^*)$ is the so-called function of organizational co-operativity expressed as

$$\Omega(\sigma_o^*) = \frac{\zeta_3\sigma_o^{*3} + \zeta_2\sigma_o^{*2} + \zeta_1\sigma_o^*}{\zeta_4'\sigma_o^{*4} + \zeta_3'\sigma_o^{*3} + \zeta_2'\sigma_o^{*2} + \zeta_1'\sigma_o^* + \zeta_o'} \tag{94}$$

The ζ and ζ' parameters are combinations of the two moments var(δ^*) and cov(δ^*,ε^*). One may find

$$\begin{aligned}
\zeta_3 &= -\operatorname{var}(\delta^*) + \operatorname{cov}^2(\delta^*, \varepsilon^*)\\
\zeta_2 &= -2\,\{\operatorname{var}(\delta^*) + \operatorname{cov}(\delta^*, \varepsilon^*)\operatorname{var}(\delta^*) + \operatorname{cov}^2(\delta^*, \varepsilon^*)\}\\
\zeta_1 &= -\operatorname{var}(\delta^*) + \{\operatorname{cov}(\delta^*, \varepsilon^*) - \operatorname{var}(\delta^*)\}^2
\end{aligned} \tag{95}$$

and

$$\begin{aligned}
\zeta_4' &= 1 + \operatorname{cov}(\delta^*, \varepsilon^*)\\
\zeta_3' &= 4 - \operatorname{var}(\delta^*) + 2\operatorname{cov}(\delta^*, \varepsilon^*) - \operatorname{cov}^2(\delta^*, \varepsilon^*)\\
\zeta_2' &= 6 - \operatorname{var}(\delta^*) - 2\operatorname{cov}^2(\delta^*, \varepsilon^*) + 2\operatorname{cov}(\delta^*, \varepsilon^*)\operatorname{var}(\delta^*)\\
\zeta_1' &= 4 + \operatorname{var}(\delta^*) - 2\operatorname{cov}(\delta^*, \varepsilon^*) - \{\operatorname{cov}(\delta^*, \varepsilon^*) - \operatorname{var}(\delta^*)\}^2\\
\zeta_o' &= 1 + \operatorname{var}(\delta^*) - \operatorname{cov}(\delta^*, \varepsilon^*)
\end{aligned} \tag{96}$$

As var(δ^*) and cov(δ^*,ε^*) are always smaller than unity, one may show that all the terms of the numerator of equation (94) are negative, whereas all the terms of the denominator of the same equation are positive. Therefore $\Omega(\sigma_o^*)$ is always negative and the co-operativity generated by the spatial organization of fixed charges and enzyme molecules can only be negative.

It has been considered above (Figure 7) that different types of organization of enzyme molecules and fixed charges may occur in the matrix. The theoretical results that have been presented so far allow one to express quantitatively how different types of organization affect the reaction rate. If the enzyme molecules and the fixed charges are homogeneously, or randomly, distributed in the matrix (Figure 7A) the equation that expresses this situation is indeed

$$v = \frac{VS_o^2}{K\Delta + S_o^2} \tag{97}$$

If the charges are clustered but the enzyme molecules randomly distributed in the matrix, the enzyme molecules that are located outside the clusters are not submitted to electrostatic repulsion effects, and thus follow Michaelis-Menten kinetics. Alternatively the enzyme molecules that are located in the charge clusters are indeed submitted to these effects. In these clusters the bivariate moments $\mu(\delta^*,\varepsilon^*)$ are all equal to zero. Therefore the rate equation is

$$v = \frac{V_1 S_o}{K + S_o} + \frac{N \langle V_2 \rangle S_o^2/K \langle \Delta \rangle}{1 + S_o^2/K \langle \Delta \rangle} \left\{ 1 + \sum_{r=1}^{m} (-1)^r \frac{\mu_r(\delta^*)}{\left(1 + S_o^2/K \langle \Delta \rangle\right)^r} \right\} \quad (98)$$

Alternatively if the charges are homogeneously distributed in the matrix whereas the enzyme molecules are clustered, the monovariate and bivariate moments of the charge distribution are equal to zero. Therefore the corresponding rate equation is

$$v = \frac{V_1 S_o}{K + S_o} + \frac{N \langle V_2 \rangle S_o^2/K \langle \Delta \rangle}{1 + S_o^2/K \langle \Delta \rangle} \quad (99)$$

In both equations (98) and (99) V_1 is proportional to the density of enzyme molecules not submitted to electrostatic effects, whereas $\langle V_2 \rangle$ is proportional to the mean enzyme density in charge or in enzyme clusters.

If the clusters of enzyme and the clusters of charges are only in part superimposed, some enzyme molecules will not be submitted to electrostatic repulsion effects and will respond hyperbolically to a change of substrate concentration whereas another population of these molecules will be submitted to these electrostatic effects. The corresponding rate equation is thus

$$v = \frac{V_1 S_o}{K + S_o} + \frac{N \langle V_2 \rangle S_o^2/K \langle \Delta \rangle}{1 + S_o^2/K \langle \Delta \rangle} \left\{ 1 + \sum_{r=1}^{m} (-1)^r \frac{\mu_r(\delta^*) + \mu_{r,1}(\delta^*,\varepsilon^*)}{\left(1 + S_o^2/K \langle \Delta \rangle\right)^r} \right\} \quad (100)$$

Last if the charge clusters are exactly superimposed to the enzyme clusters, the corresponding rate equation is indeed equation (89).

6. An example of enzyme behaviour in organized biological systems: the dynamics of enzymes bound to plant cell walls

An acid phosphatase is ionically bound to plant cell walls. (Crasnier et al. 1980, 1985; Noat et al. 1980, Darvill et al. 1981). This enzyme may be solubilized and purified to homogeneity. It is a monomeric glycoprotein of about 100 000 molecular mass. The enzyme in free solution displays classical Michaelis-Menten kinetics. At rather "high" ionic strength the cell wall-bound phosphatase follows the same kinetics. At "low" ionic strength however it shows a complex, mixed, co-operativity. Moreover at constant substrate concentration, the enzyme activity in free solution is, to a large extent, independent upon the ionic strength, whereas the activity increases with the ionic strength if the enzyme is bound to cell walls (Figure 9). As the substrate of the enzyme is a phosphate, there must be electrostatic repulsions of this substrate by the fixed negative charges of the cell wall. Therefore the apparent activation of the bound-enzyme at high ionic strength must be due to the suppression of this electrostatic repulsion under these conditions. As a matter of fact, the kinetics of the bound phosphatase at "low" ionic strength is best fitted with equations (98) and (99). This may imply that the charges are homogeneously distributed in the matrix whereas the enzyme molecules are clustered (equation 98). Another possible explanation is that both enzyme molecules and fixed charges are clustered but that the variance of this charge distribution is close to zero (equation 99). As the first interpretation is simpler and more likely than the second one it has to be retained, at least as a provisional approximation of reality (Dussert et al. 1989).

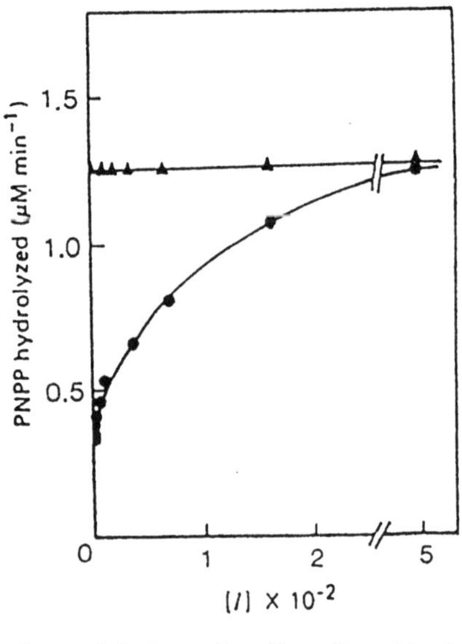

Fig. 9. Ionic strength modulation of cell wall acid phosphatase activity. The substrate concentration is held constant whereas the ionic strength is varied. The bound enzyme appears "activated" by ionic strength (●) whereas the "soluble" enzyme is not (▲) (from Noat et al. 1980)

The view that enzyme molecules are clustered may be directly confirmed by image analysis. The degree of spatial order, or disorder, that may exist among a population of material points in a plane may be quantitatively expressed through the use of the so-called minimal spanning tree (Dussert et al. 1987). A minimal spanning tree is a connected graph obtained by joining each point to its nearest neighbour. The distribution of the edge lengths of all the possible graphs pertaining to a set of points follows a Laplace-Gauss law. The mean (m) and the standard deviation (σ) of this distribution may be scaled in order to allow their comparison to the corresponding parameters pertaining to a different set of points. For randomly scattered points the scaled mean and standard deviation assume the values 0.662 and 0.311, respectively. In the (σ,m) plane, any deviation from these values, that express absolute disorder, implies that some sort of order exists in the spatial distribution of the points (Figure 13). This method may be applied to quite different types of material points, from stars in the sky, to enzyme molecules on an electron micrograph.. The acid phosphatase molecules of plant cell walls display deviations of m and σ values that are significantly different from a spatial random distribution and indicate some sort of enzyme clustering (Figure 10).

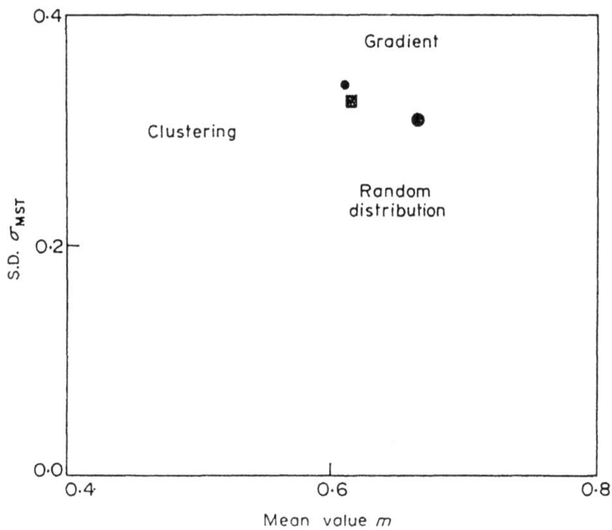

Fig. 10. The (m, σ_{MST}) plane. The "large" dot (●) pertains to a perfectly random distribution. If the mean of the experimental data, m, is below that of a random distribution, this is indicative of a trend towards clustering. If the standard deviation of the data is above that of a random distribution there is a trend towards the formation of a gradient. The "small" dot (•) and the square (■) pertain to the distribution of acid phosphatase in sycamore cell walls obtained for different experimental conditions (from Ricard et al. 1992)

It is well known that calcium may be reversibly bound to plant cell walls and that this binding process may result in the inhibition of cell wall extension (Aspinall 1980; Cleland and Rayle 1977; Crasnier et al. 1985; Lamɪ 1980; Taiz 1984). As this process is in part due to the activity of enzymes, one may wonder whether the effect of calcium on growth primarily be exerted on the cell wall electric field that affects enzyme activity. The idea that calcium may alter the activity of cell wall-bound enzymes may be demonstrated by the following experiment. If calcium is taken away from cell wall fragments by acidic treatment, the apparent activity of bound acid phosphatase is markedly decreased and the enzyme displays a strong negative co-operativity. Loading the cell wall fragments with calcium results in an apparent activation of the bound enzyme together with a decrease of the negative co-operativity. This treatment may even result in the complete suppression of co-operativity together with a strong activation of the enzyme (Figure 11). Moreover calcium has no effect on the enzyme in free solution. These results are understandable on the basis of the view that calcium binding results in a decrease of the net fixed charge density Δ^- of the cell wall. This in turn implies that the $\Delta\psi$ value of the cell wall declines, thus resulting in a decrease of this electrostatic repulsion of the phosphate (the substrate of the phosphatase) by the cell wall and therefore an apparent activation of the enzyme.

These ideas may be applied to the understanding, in physico-chemical terms, of plant cell wall extension . Primary plant cell wall is made up of cellulose microfibrils interconnected by acidic polysaccharides. The fixed charge density of the wall is due to these components of the cell envelope. Plant cell wall growth requires the sliding of cellulose microfibrils under the influence of turgor pressure. This process of microfibril sliding most probably involves endoglycosyl transferases that catalyse the breaking of β $(1 \rightarrow 4)$ bonds that crosslink cellulose microfibrils. This process is also associated with the incorporation of new polysaccharide units in the wall. Pectins are incorporated in a neutral, methylated, form. If these were the only biochemical processes associated with cell wall growth, one would have expected the fixed charge density to continuously decline as the cell extends. The cell wall however possesses an enzyme, a pectin methyl esterase, that demethylates methylated pectin and therefore generates the fixed negative charges of the wall required to hold its electrostatic potential at a fixed value (Moustacas et al. 1986; Nari et al. 1986; Ricard and Noat 1986). This succession of biochemical events: incorporation of methylated precursors in the wall, sliding of cellulose microfibrils, and restoration of the initial fixed charge density through pectin methyl esterase is shown in the hypercycle of Figure 12. This hypercycle is formally equivalent to two antagonistic enzyme reactions: one which results in a decrease of the fixed charge density (the reactions associated with cell wall growth) and another one which results in an increase of this density (the pectin methyl esterase reaction). All the glucanases that are involved in cell wall metabolism have an acidic optimum pH, between 4 and 5. Alternatively pectin methyl esterase has an optimum pH of about 7.5. The difference of pH-sensitivity of these two types of reactions must result in a steep transition of the charge density of the wall as a function of local, or bulk, pH (Figure 12) (Goldbeter and Koshland 1981, 1984).

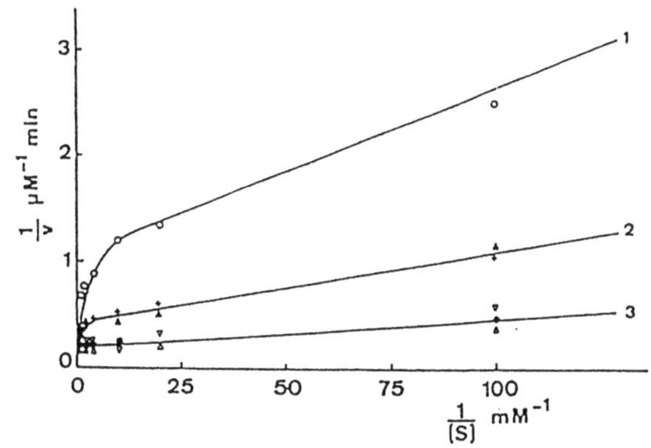

Fig. 11. Apparent negative co-operativity of plant cell wall-bound acid phosphatase at low ionic strength. Curve *1* shows the apparent negative co-operativity of acid phosphatase bound to calcium-deprived cell walls. Curve *2* shows the co-operativity of the same enzyme bound to native cell walls (+) or to calcium-deprived cell walls supplemented with calcium (▲). Curve *3* shows the suppression of apparent co-operativity of the bound enzyme after addition of 0.2 M NaCl. The same results are obtained with "native" cell walls (∇), "calcium deprived" cell walls (Δ) or "calcium deprived" cell walls supplemented with calcium (●) (from Crasnier et al. 1985)

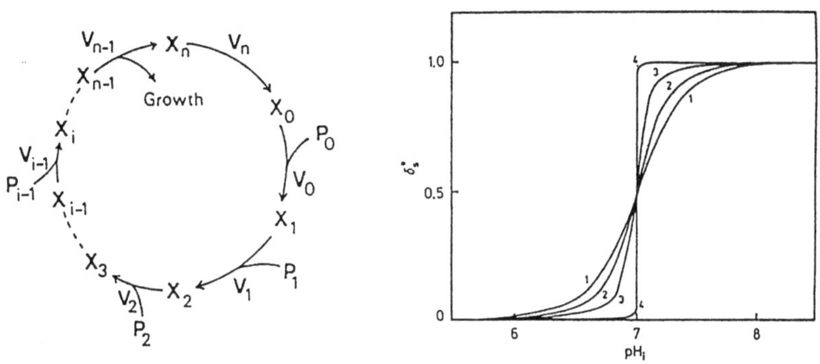

Fig. 12. The hypercycle of cell extension. In the hypercycle (left-hand side panel) X_0, ..., X_n represent the reaction intermediates, P_0, P_1, ... the methylated precursors of cell wall polysaccharides. The step $X_{n-1} \rightarrow X_n$ is that of cell wall extension and the step $X_n \rightarrow X_0$ that of methylated pectin hydrolysis. The curves *1 - 4* of this Fig. show the sharp response that may be expected from this hypercycle. δ_s^* is the normalized charge density, pH_i the local pH. Curves *1 - 4* are obtained for different conditions (from Ricard and Noat 1986a)

The interest of this model is that it allows a number of predictions that may be tested experimentally. The first prediction is that pectin methyl esterase should be controlled by protons and cations; the second is that this enzyme is effectively involved in the $\Delta\psi$ of the cell wall; the third is the existence of an ionic control of wall loosening enzymes; the last prediction is the existence of a strong positive co-operativity of the response of pectin methyl esterase system to slight changes of local pH.

There is little doubt that cell wall pectin methyl esterase is controlled by cations. In the absence of cations, for instance Na^+ or Ca^{2+}, pectin methyl esterase is totally devoid of activity (Goldberg 1984; Goldberg and Prat 1982). In the presence of low cation concentrations the reciprocal plots with respect to pectin are parallel. They tend to converge, however, at the same point when the cation concentration is increased (Figure 13). This implies that the pectin methyl esterase reaction is inhibited by an excess of metal (Nari et al. 1991 a).

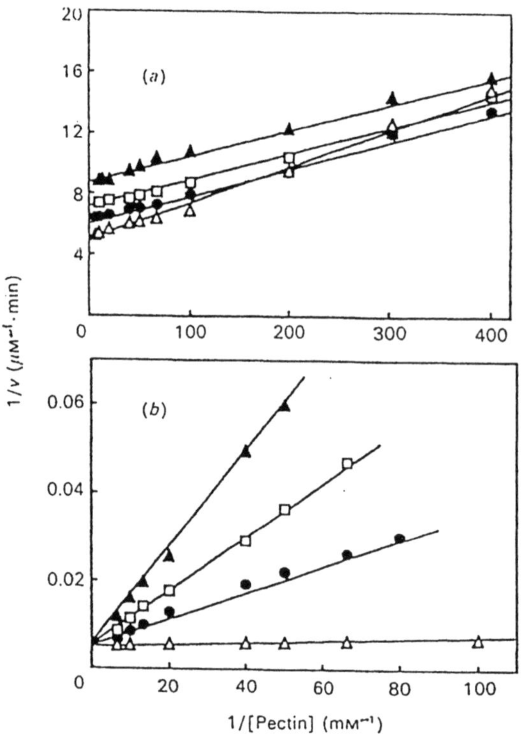

Fig. 13. Kinetics of pectin methyl esterase activity for different cation concentration. **a** Kinetics at "low" sodium concentration. **b** Kinetics at "high" sodium concentration. (from Nari et al. 1991)

These effects may be mimicked to a large extent by methylene blue that may be stacked to the polyanion. In the absence of NaCl, methylene blue may "activate" pectin methyl esterase.

These results, as well as others, suggest that the metal does not interact with the enzyme but with pectin. Numerous studies have shown that there exist, in natural pectins, "blocks" of carboxylate groups. These blocks tend to trap enzyme molecules which are then not available for catalysis. In the presence of metal ions, or of methylene blue, the negative charges of the polyanion are neutralized and the enzyme molecules are released from the "blocks". The binding of metal ions to the "blocks" is a non co-operative process. Moreover it is well known that a methylated unit can undergo demethylation only if the neighbouring units are demethylated and thus neutralization of negatively charged groups by metal ions should therefore result in a decrease of the reaction rate. The inhibition by excess of metal which has already been mentioned is most probably to be ascribed to the binding of metal ions to these carboxyl groups. Kinetic results suggest this second type of binding process to be co-operative.

The tentative reaction mechanism of pectin hydrolysis is shown in Figure 14. The corresponding kinetic model that accommodates the results above is shown in Figure 15.

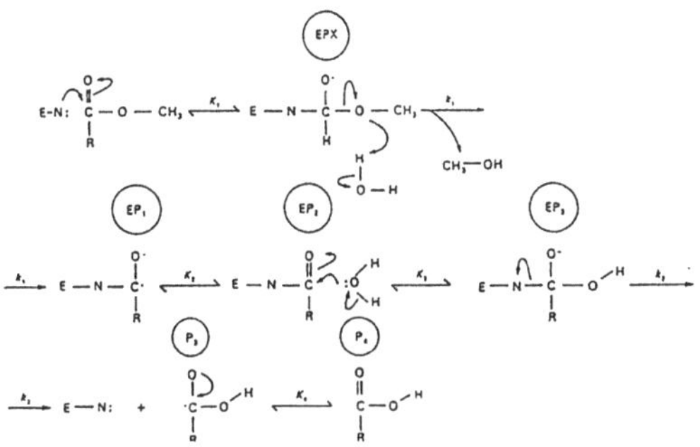

Fig. 14. Tentative mechanism of action of pectin methyl esterase. The symbols EPX, EP_1, EP_2, EP_3, P_2 and P_3 also appear in the next Figure

The rate equation which may be derived from this model fits exactly the rate measurements. As outlined previously, this model is based on several important ideas:

- the metal does not interact with the enzyme but with the polyanion;
- the binding of the metal to the "blocks" of carboxyl groups brings about the release of the enzyme trapped to these "blocks";
- the binding of metal ions to the carboxyl groups adjacent to the ester bonds to be cleaved results in an inhibition of the reaction;
- the second process is co-operative whereas the first is not.

Another important idea which is implicit in the hypercycle of Figure 12 is that pectin methyl esterase is involved, *in vivo*, in the building up of the electrostatic potential of the cell wall. This idea may be tested experimentally. When the ionic strength of a suspension of cell wall fragments is raised one may monitor an efflux of protons in the bulk phase. This is due to the fact that the fixed negative charges of the cell wall tend to attract protons at low ionic strength. However as the ionic strength is raised in the bulk phase, protons that were initially attracted in the polyanionic matrix tend to diffuse outside that matrix and may be titrated. If one knows the volume of the bulk phase and that of the cell wall, one may estimate, from the extent of proton efflux, the local proton concentration in the cell wall as well as the corresponding $\Delta\psi$ value.

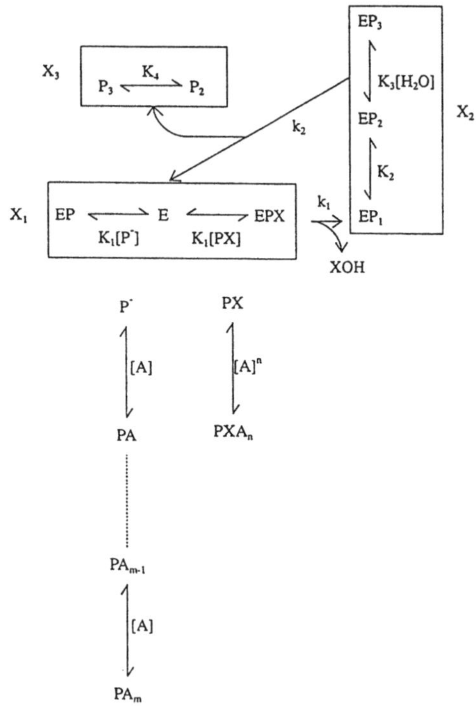

Fig. 15. Kinetic model of pectin methyl esterase activity. The steps represented in the boxes X_1, X_2, X_3 are assumed to be in fast equilibrium. In the box X_1 the enzyme is assumed to interact either with a free (P^-) or with a methylated (PX) carboxyl group. Metal ions (A) may be bound either non-co-operatively to the "blocks" of carboxyl groups (formation of PA to PA_m) or co-operatively to the carboxyl groups adjacent to the methyl group to be cleaved (formation of PXA_n). The first process results in the release of the free active enzyme initially trapped to the "blocks" and the second results in an inhibition of the enzyme reaction. The significance of the intermediates, EPX, EP_1, EP_2, ... is to be found in Fig. 14 (from Nari et al. 1991)

Since cell wall-bound pectin methyl esterase is activated at "high" pH and at "high" metal ion concentration, one should expect that, after pre-treatment of the walls at "high" pH or ionic strength, the $\Delta\psi$ of this organelle be increased. This is precisely what is observed. Cell wall fragments are prepared at pH 8. A first sample is pre-incubated 15 minutes at the same pH (control) and the proton efflux observed after raising the ionic strength is measured under these conditions of pH. Other samples are pre-incubated for the same time at higher pH values, then washed and transferred at pH 5. The proton efflux is then measured under these conditions. One may observe that the extent of proton efflux increases as the pH of pre-incubation is increased. From the extent of the proton efflux one may estimate the local proton concentration in the cell wall and the corresponding $\Delta\psi$ value. These values are plotted against the pH of the pre-incubation mixture (Figure 16). These results leave no doubt as to the part played by pectin methyl esterase on the building up of the cell wall electrostatic potential *in vivo*. (Moustacas et al. 1986, Nari et al 1986).

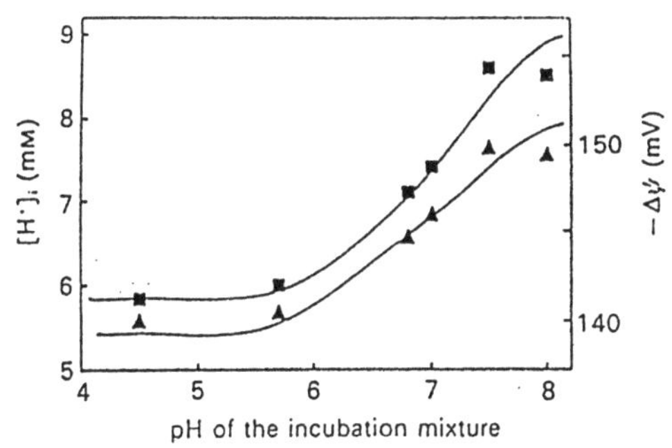

Fig. 16. Variation of the local proton concentration (■) and of the $\Delta\Psi$ of the cell wall (▲) as a function of the pH of the incubation mixture (see text)

Moreover, if pectin methyl esterase were effectively involved in the building up of this potential, one should expect that the same concentration of metal that results in the activation of the enzyme produces an increase of the electrostatic potential. This again has been observed (Figure 17), and cannot be fortuitous (Moustacas et al. 1991).

The theoretical model discussed above implies that $\Delta\psi$ is the trigger of growth and that the overall activity of glucanases involved in plant cell growth depends upon the ionic atmosphere of the wall and therefore upon its electrostatic potential. This idea may be confirmed experimentally. Cell wall extension is accompanied by a limited cell wall autolysis, which can be monitored by the release from the wall of reducing sugars. This release may be considered as a measure of the activity of wall loosening enzymes involved in cell growth. These enzymes have an optimum pH which is about 5. One may also wonder whether metal ion concentration alters the activity of these bound enzymes. It appears to be so. If the rate of autolysis is monitored at pH 5 but at "low" (0.025) or "high" (0.125) ionic strength, this rate is higher at "high" than at "low" ionic strength. Moreover, if the cell wall fragments are incubated 15 minutes at pH 8, then washed, and if their autolytic activity is monitored at pH 5 and at "low" (0.025) ionic strength, the rate is higher than the one observed in the absence of any pre-incubation at pH 8. As previously outlined, this pre-incubation stimulates pectin methyl esterase, which in turn results in an increased charge density. These fixed negative charges attract cations and this stimulates the activity of wall loosening enzymes.

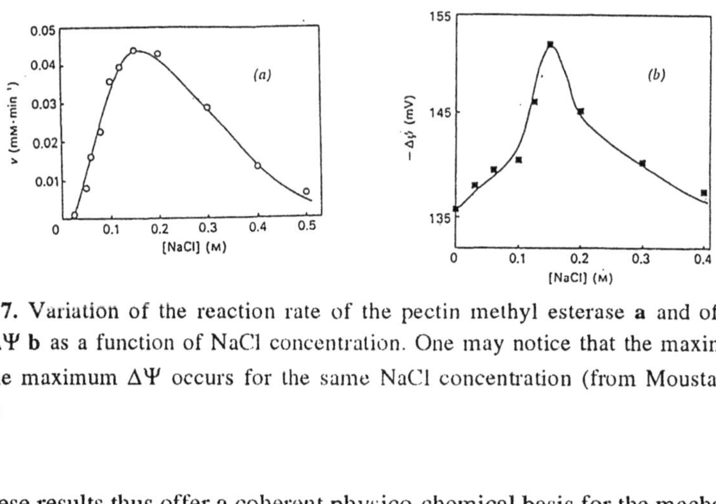

Fig. 17. Variation of the reaction rate of the pectin methyl esterase **a** and of the cell wall $\Delta\Psi$ **b** as a function of NaCl concentration. One may notice that the maximum rate and the maximum $\Delta\Psi$ occurs for the same NaCl concentration (from Moustacas et al 1991)

These results thus offer a coherent physico-chemical basis for the mechanism of plant cell wall growth. When the $\Delta\psi$ value of the wall is large, the local proton concentration is large as well. Pectin methyl esterase is thus inactive and wall loosening enzymes display a strong activity. The cell wall thus extends under the influence of the turgor pressure. But this extension in turn results in the decrease of the electrostatic potential of the wall. This decrease is due to the fact that pectins are incorporated in a methylated state. As the wall extends, the density (but not the number) of the fixed negative charges declines, thus giving rise to the decrease of $\Delta\psi$ values, which in turn results in an increase of the local pH in the wall. This alteration of the local pH values brings about an activation of pectin methyl esterase which restores the initial $\Delta\psi$ value. This scheme however embodies one conceptual difficulty. This difficulty is that proton and metal ions affect pectin methyl esterase in two adverse ways. Namely, high proton concentrations result in an inhibition of pectin methyl esterase whereas high metal ion concentrations bring about an apparent activation of this enzyme. Moreover, as a change of the fixed charge density in the wall should result in the simultaneous increase or decrease of both metal ion and proton concentrations, one should expect the effect of these two changes to be mutually antagonistic.

In fact this difficulty is only apparent, as one may observe that the metal concentration that gives a maximum reaction rate of pectin methyl esterase is much lower under alkaline than under acidic pH conditions (Figure 18). This has two implications : first, after the cell has extended, at "low" $\Delta\psi$ values, the simultaneous "low" metal ion and "low" proton concentrations should result in a "high" pectin methyl esterase activity; second, when the $\Delta\psi$ values increase, "high" proton and metal ion concentrations may still result in a significant pectin methyl esterase activity and therefore in an amplification of the increase of $\Delta\psi$.

The model, which has been presented as an hypercycle in Figure 9A, implies that the build up of cell wall charge density is co-operative relative to the local pH changes (Figure 9B). As the cell wall extends, the local proton and metal ion concentration declines and this triggers pectin methyl esterase activity, which restores the initial $\Delta\psi$ value. If this restoration were not co-operative, the electrostatic potential of the wall would remain constant during cell growth. If alternatively pectin methyl esterase did not respond, or poorly responded, to changes of metal and proton concentration, the fixed charge density should decline during cell wall extension. Last, if the response of pectin methyl esterase were co-operative, one should observe altogether an increase of the charge density and of the cell wall volume. This prediction may be tested experimentally.

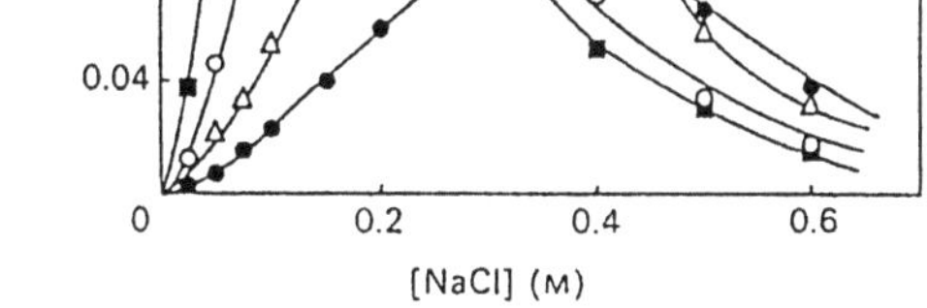

Fig. 18. Variation of the reaction rate of pectin methyl esterase as a function of NaCl concentration and for different pHs. As the pH increases (from 5 to 8) the maximum reaction velocity increases and is obtained for lower NaCl values (from Moustacas et al. 1991)

Isolated clumps of plant cells in sterile culture represent a biological system that allows one to study cell growth. When clumps of sycamore or soybean cells are transferred to a fresh culture medium, the cells extend first in a more or less synchronous way, then divide. A possible way of estimating the fixed charge density of the cell wall is to measure their content in unmethylated uronic acids. If the percent of uronic acids is plotted as a function of time, together with the cell wall volume, the percent of uronic acids increases at first, then decreases when the cell extends. These results thus show that the response of pectin methyl esterase to changes of proton and metal ion concentration is co-operative.

The scheme of Figure 19 summarizes the main conclusions that have been presented above. The trigger of growth is the $\Delta\psi$ value. When the electrostatic potential of the cell wall is low, the local proton and metal ion concentration is low as well, and this brings about an activation of pectin methyl esterase (loops 1 and 2) which tends to generate higher $\Delta\psi$ value. This, in turn, results in the attraction of protons and metal ions in the polyanionic matrix, and this tends to amplify the increase of the electrostatic potential. This amplification is the consequence of two physical processes: the shift of the reaction profile towards "high" metal ion concentration when the pH drops (Figure 18); the increase of availability of pectin methyl esterase which tends to dissociate from the "blocks" of carboxyl groups as the metal ion concentration increases. This brings about an optimal proton and metal ion concentration for the activity of glucanases involved in cell wall extension. If the metal ion (and the proton) concentration in the wall were too high, the fine tuning of this complex process would be achieved through the apparent inhibition of pectin methyl esterase by an excess of the metal. As this inhibition is exerted through the binding of metal ions to carboxyl groups of the polyanion, the $\Delta\psi$ value of the wall decreases as well (loop 3), (Ricard 1986 a, 1987 a and b, Moustacas et al 1991).

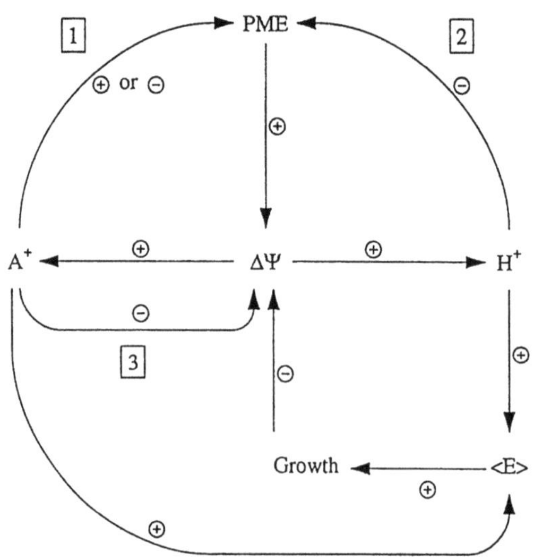

Fig. 19. The control of cell-wall growth and pectin methyl esterase activity by cell-wall $\Delta\Psi$. The ⊕ and ⊖ signs stand for activation and inhibition, respectively (see text)

7. Control and dynamics of enzyme networks

It has been considered so far how the interactions between an enzyme and its environment may control the catalytic activity. There exists another type of organization where a reactant is the product of an enzyme reaction and the substrate of another one. From that point of view a metabolic network may be considered as an organized set of enzyme reactions. The very property that these enzyme reactions are linked confers upon the whole system properties that are novel with respect to those of individual enzyme reactions. Indeed this type of organization is not exclusive of any other type of interaction that may exist between the enzymes and their environment. In this section we are going to consider the control of a linear metabolic network in steady state, the dynamics of a simple metabolic cycle, and the dynamics of a metabolic cycle at the surface of a charged membrane.

7.1. Control of a linear metabolic network

Let there be the linear sequence of enzyme reactions

$$S_o \underset{v_1}{\overset{E_1}{\longrightarrow}} S_1 \underset{v_2}{\overset{E_2}{\longrightarrow}} S_2 \; \ldots \; S_{n-1} \underset{v_n}{\overset{E_n}{\longrightarrow}} S_n$$

where $S_1, .., S_{n-1}$ are the intermediates of the metabolic pathway, $E_1, .., E_n$ the corresponding enzymes and $v_1, .., v_n$ the individual reactions rates. The flux of this pathway is defined as

$$J = \frac{dS_n}{dt} \tag{101}$$

According to the classical views of metabolic control, that were developed between 1960 and 1965, it is classically considered that this flux is regulated thanks to an allosteric enzyme that is retro-inhibited or activated by one of the last intermediates of the metabolic network. Although this might be true in a number of cases, this is not necessarily true for all the cases. The aim of the so-called metabolic control theory is precisely to discover the general laws that control the steady state behaviour of metabolic pathways. Control of a metabolic flux is effected thanks to two types of parameters: the control and the elasticity coefficients (Fell and Sauro 1985; Heinrich and Rapoport 1974; Heinrich et al. 1977; Kacser 1987; Kacser and Burns 1979; Westerhoff and Van Damm 1987; Giersch 1988a, b).

Control coefficients

The control coefficients of the linear sequence of reactions are defined as

$$C_i = \frac{E_i}{J}\frac{\partial J}{\partial E_i} = \frac{\partial \ln J}{\partial \ln E_i} \qquad (i = 1, ..., n) \qquad (102)$$

They express how an enzyme concentration, or an enzyme reaction rate (since rates are proportional to enzyme concentrations) affects the overall metabolic flux. These parameters, which have been proposed independently by Kacser and Burns and Heinrich and Rapoport, are systemic properties since they refer to the behaviour of the overall metabolic pathway.

Whatever the kinetic properties of the individual enzymes of the pathway , one has

$$J = f(E_1, E_2, ..., E_n) \qquad (103)$$

Moreover the flux may be considered as a state function, therefore

$$dJ = \sum_{i=1}^{n} \left(\frac{\partial J}{\partial E_i}\right) dE_i \qquad (104)$$

Dividing both sides by J, rearranging and taking account of the definition of control coefficients yields

$$\frac{dJ}{J} = \sum_{i=1}^{n} C_i \frac{dE_i}{E_i} \qquad (105)$$

This expression shows that the relative perturbation of the flux is a linear combination of the perturbations of the individual enzymes.

There are two different ways of controlling a sequence of enzyme reactions and, at the same time, to maintain the steady state unchanged. When an enzyme is perturbed through an external parameter the first possibility offered to the system to hold its steady state unaffected, is to change in the same way the properties of all the enzymes involved in that pathway. The second possibility of controlling the reaction sequence and to keep the steady state unaltered, is to modify, in a concerted way, the properties of the individual enzymes so as to annihilate the effect of the perturbation which has been exerted upon one of the enzyme reactions. The first strategy of control leads to the property of summation, whereas the second one leads to the property of connectivity.

The property of summation is quantitatively expressed by the so-called summation theorem. The property of connectivity is associated with the definition of elasticities that will be considered below. If the concentration, or the activity, of all the enzymes of a reaction network is changed by the same value α, namely if

$$\frac{dE_i}{E_i} = \alpha \qquad (i = 1, ..., n) \qquad (106)$$

then the steady state will remain unchanged and the reaction flux will be varied by the same value

$$\frac{dJ}{J} = \alpha \tag{107}$$

Inserting expressions (106) and (107) into equation (105) yields

$$\sum_{i=1}^{n} C_i = 1 \tag{108}$$

which is the mathematical expression of the summation theorem. This theorem states that if a system is strictly controlled, that is if its steady state is unchanged despite a perturbation of a flux, the sum of the control coefficients cannot all be independent and their sum must be equal to unity. This means that different enzymes of the same pathway, and possibly all the enzymes of that pathway, may contribute to the control of the reaction flux. As previously outlined, this conclusion is at variance with the assumption commonly made that a metabolic pathway is controlled through a specific enzyme sensitive to an end product of the reaction chain.

From a mathematical point of view, the summation theorem is equivalent to the statement that a flux J is a homogeneous function of degree one in enzyme concentrations $E_1, E_2, ..$ A function $f(x_1, x_2, ..)$ is called homogeneous of degree h if $f(tx_1, tx_2, ..) == t^h f(x_1, x_2)$ for all the $t \neq 0$. The Euler's theorem establishes that a function $f(x_1, x_2, ..)$ is homogeneous of degree h in $x_1, x_2, ..$ if

$$hf(x_1, x_2, ...) = \sum_j x_j \frac{\partial f}{\partial x_j} \tag{109}$$

Conversely, any function $g(x_1, x_2, ..)$ that obeys the relationship

$$hg(x_1, x_2, ...) = \sum_j x_j \frac{\partial g}{\partial x_j} \tag{110}$$

is homogeneous of degree h in $x_1, x_2, ..$ As the expression of the summation theorem (equation 108) may be rewritten as

$$\sum_i E_i \frac{\partial J}{\partial E_i} = J \tag{111}$$

which is formally equivalent to equations (109) and (110) if h = 1, the conclusion of this reasoning is thus that the summation theorem is equivalent to the proposition that the overall metabolic flux is a homogeneous function of degree one in enzyme concentrations. Moreover if a reaction rate is a homogeneous function of degree one in enzyme concentration, this implies that the reaction rate is proportional to the enzyme concentration. Thus if

$$t\, v_i\, (E_i) = v_i\, (t\, E_i) \tag{112}$$

this means that changing the enzyme concentration by a factor t, changes the rate by the same factor. In other words the rate must be proportional to the enzyme concentration.

Elasticity coefficients

The control coefficients have been defined as systemic parameters expressing how an enzyme concentration (or an enzyme reaction rate) affects the overall reaction flux. The second type of coefficient which is required to define the mechanism of control of a steady state is called an elasticity coefficient. This type of coefficient expresses how an enzyme concentration, E_i, and different reaction intermediates, S_i, .., S_j, affect a given enzyme reaction velocity, v_i, within the overall reaction pathway. This reaction velocity, v_i, is indeed a function of the corresponding enzyme concentration E_i and of a number of reactants, S_i, .., S_j, .., that take part in, or control, this process

$$v_i = f\,(E_i, S_i, ..., S_j, ...) \tag{113}$$

For a linear sequence of enzyme reactions, these coefficients are expressed as

$$\varepsilon_j^i = \frac{S_j}{v_i}\frac{\partial v_i}{\partial S_j} = \frac{\partial \ln v_i}{\partial \ln S_j} \tag{114}$$

These coefficients express how an intermediate S_j affects the reaction rate, v_i, of the overall metabolic pathway. Moreover the total differential of the reaction rate is

$$d\, v_i = \frac{\partial v_i}{\partial E_i} dE_i + \sum_j \frac{\partial v_i}{\partial S_j} dS_j \tag{115}$$

This expression may be rewritten as

$$d\, v_i = \frac{\partial v_i}{\partial E_i}\frac{E_i}{v_i}\frac{dE_i}{E_i} v_i + \sum_j \frac{\partial v_i}{\partial S_j}\frac{S_j}{v_i}\frac{dS_j}{S_j} v_i \tag{116}$$

Each enzyme reaction rate of the pathway may be considered as a homogeneous function of the corresponding enzyme concentration. Therefore

$$\frac{E_i}{v_i}\frac{\partial v_i}{\partial E_i} = 1 \tag{117}$$

and equation (116) may be rewritten as

$$dv_i = \frac{d E_i}{E_i} v_i + \sum_j \varepsilon_j^i \frac{d S_j}{S_j} v_i \tag{118}$$

or

$$\frac{dv_i}{v_i} = \frac{d E_i}{E_i} + \sum_j \varepsilon_j^i \frac{d S_j}{S_j} \tag{119}$$

This expression allows justifing the term "elasticity" given to this parameter. It shows that if the concentration of the enzyme is perturbed, the corresponding rate, v_i, and therefore the overall reaction flux, may well not be perturbed if

$$\frac{dE_i}{E_i} = -\sum_j \varepsilon_j^i \frac{d S_j}{S_j} \tag{120}$$

It is therefore because the system displays an "elasticity" that a perturbation of an enzyme concentration may well not perturb the flux.

This equation represents the condition required to hold the flux constant upon changing simultaneously the concentrations of the intermediates by the values dS_1, dS_2, .., dS_j. If the concentration of only one intermediate S_j, is altered, then

$$\frac{dE_i}{E_i} = -\varepsilon_j^i \frac{d S_j}{S_j} \tag{121}$$

under this condition, equation (105) becomes

$$\frac{dJ}{J} = \sum_{i=1}^{n} C_i \frac{dE_i}{E_i} = 0 \tag{122}$$

and inserting expression (121) into this equation (122) yields

$$\frac{dJ}{J} = 0 = \sum_{i=1}^{n} C_i \varepsilon_j^i \frac{dS_j}{S_j} = \frac{dS_j}{S_j} \sum_{i=1}^{n} C_i \varepsilon_j^i \tag{123}$$

As

$$\frac{dS_j}{S_j} \neq 0 \tag{124}$$

the condition for dJ/J = 0 is of necessity

$$\sum_{i=1}^{n} C_i \varepsilon_j^i = 0 \tag{125}$$

and this equation expresses the property of connectivity. In the case of a linear reaction sequence there exists n enzymes (or n enzyme reactions) and n-1 reaction intermediates. This property of connectivity expresses how one of the reaction intermediates affects the n enzymes and one has thus

$$\sum_{j=1}^{n-1} \sum_{i=1}^{n} C_i \varepsilon_j^i = 0 \tag{126}$$

The property of connectivity may be associated with the property of summation in the matrix relation

$$\begin{bmatrix} 1 \\ 0 \\ 0 \\ \cdots \\ 0 \end{bmatrix} = \begin{bmatrix} 1 & 1 & \cdots & 1 \\ \varepsilon_1^1 & \varepsilon_1^2 & \cdots & \varepsilon_1^n \\ \varepsilon_2^1 & \varepsilon_2^2 & \cdots & \varepsilon_2^n \\ \cdots & \cdots & \cdots & \cdots \\ \varepsilon_{n-1}^1 & \varepsilon_{n-1}^2 & \cdots & \varepsilon_{n-1}^n \end{bmatrix} \begin{bmatrix} C_1 \\ C_2 \\ C_3 \\ \cdots \\ C_n \end{bmatrix} \tag{127}$$

The general conclusion that may be drawn from the above reasoning is that, owing to the coupling of the enzyme reactions, the overall properties of the system are indeed different from those of the individual enzymes. This theoretical approach has been found useful for the experimental study of the control of metabolic pathways.

7.2. Dynamic organization of a metabolic cycle in homogeneous phase

The simplest metabolic cycle that may be conceived is made up of two antagonistic enzyme reactions, an input and an output of matter (Boiteux et al. 1980; Hess et al. 1986; Ricard and Soulié 1982). There exists in cell metabolism a number of these cyclic process that are termed futile cycles. These cycles may be considered as simple models of more complex metabolic cycles (Figure 20). Some of these cycles may generate an amplified reponse to small changes of signal intensity (Chock et al 1980 a and b, Reich and Sel'kov 1981).

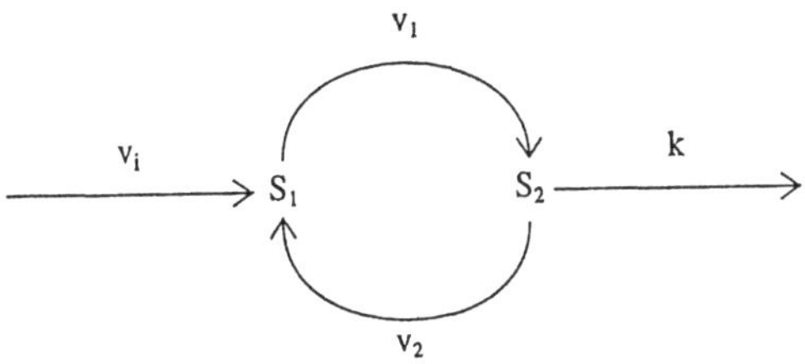

Fig. 20. A simple open metabolic cycle composed of two antagonistic enzyme reactions

If S_1 and S_2 are the concentrations of the two metabolites, one has

$$\frac{dS_1}{dt} = \frac{dS_{1\sigma}}{dt} + \frac{dx_1}{dt} = \frac{dx_1}{dt}$$
$$\frac{dS_2}{dt} = \frac{dS_{2\sigma}}{dt} + \frac{dx_2}{dt} = \frac{dx_2}{dt} \tag{128}$$

where $S_{1\sigma}$ and $S_{2\sigma}$ are the steady state concentration of S_1 and S_2 and x_1 and x_2, deviations about these steady states. If these deviations are small, the dynamic and stability analysis of this cycle may be effected thanks to the so-called phase plane technique (Nicolis and Prigogine 1977; Pavlidis 1973). Under steady state conditions one has

$$\frac{dS_{1\sigma}}{dt} = 0 = f\left(S_{1\sigma}, S_{2\sigma}\right) = v_i + v_2 - v_1$$
$$\frac{dS_{2\sigma}}{dt} = 0 = g\left(S_{1\sigma}, S_{2\sigma}\right) = v_1 - \left(v_2 + k\, S_{2\sigma}\right) \tag{129}$$

If the deviations, x_1 and x_2 from the steady state are small enough, one may expand in Taylor series these small deviations and take only the linear terms. One thus finds

$$\frac{d}{dt}\begin{bmatrix} x_1 \\ x_2 \end{bmatrix} = \begin{bmatrix} \frac{\partial f}{\partial s_1} & \frac{\partial f}{\partial s_2} \\ \frac{\partial g}{\partial s_1} & \frac{\partial g}{\partial s_2} \end{bmatrix} \begin{bmatrix} x_1 \\ x_2 \end{bmatrix} \tag{130}$$

The characteristic equation of this variational system is then

$$D^2 - T_j D + \Delta_j = 0 \tag{131}$$

where D, T_j and Δ_j are the differential operator d/dt, the trace and the determinant, respectively, of the jacobian matrix. The general solution of this system is thus

$$\begin{aligned} x_1 &= c_{11}\, e^{\lambda_1 t} + c_{12} e^{\lambda_2 t} \\ x_2 &= c_{21}\, e^{\lambda_1 t} + c_{22} e^{\lambda_2 t} \end{aligned} \tag{132}$$

where c_{11}, c_{12}, c_{21} and c_{22} are integration constants, λ_1 and λ_2 the time constants of the system that is the roots of the characteristic equation (131). As the trace and the determinant of the jacobian are equal to

$$\begin{aligned} T_j &= \lambda_1 + \lambda_2 \\ \Delta_j &= \lambda_1 \lambda_2 \end{aligned} \tag{133}$$

the dynamic behaviour of the system is defined by the respective values of T_j, Δ_j and $T_j^2 - 4\Delta_j$. The parabola

$$T_j^2 - 4\Delta_j = 0 \tag{134}$$

defines six regions in the (T_j,Δ_j) plane (Figure 21).

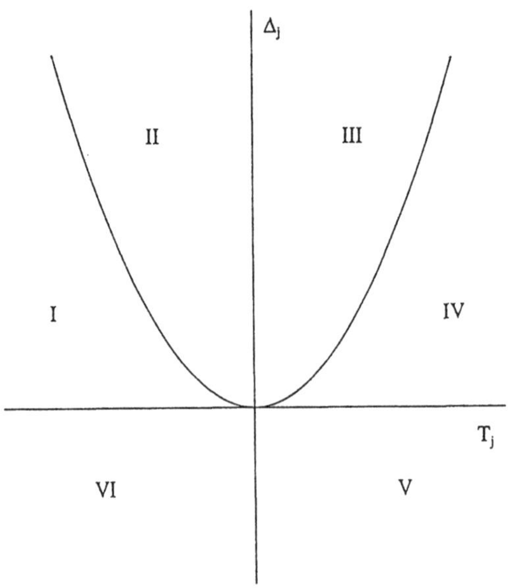

Fig. 21. The $T_j - \Delta_j$ plane

In region I one has

$$T_j < 0, \qquad \Delta_j > 0, \qquad T_j^2 > 4\Delta_j \tag{135}$$

The two roots λ_1 and λ_2 are thus real and negative. This means that the system is stable. If perturbed from its initial steady state, it monotonically returns back to the same steady state (Figure 22). In region IV

$$T_j > 0, \qquad \Delta_j > 0, \qquad T_j^2 > 4\Delta_j \tag{136}$$

the two roots are thus real and positive, and the system is unstable. When perturbed from its initial steady state it drifts monotonically to a new steady state (Figure 22).

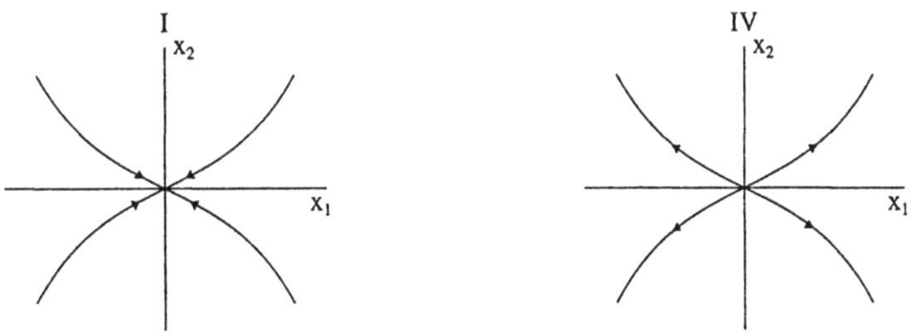

Fig. 22. Stable and unstable node of a metabolic cycle

In regions II and III, the situation is completely different. For region II one has

$$T_j < 0, \qquad \Delta_j > 0, \qquad 4\Delta_j > T_j^2 \tag{137}$$

which implies that the two roots are complex with a negative real part, namely

$$\lambda_{1,2} = \frac{T_j}{2} \pm i\omega \tag{138}$$

with

$$\omega = \frac{1}{2}\sqrt{4\Delta_j - T_j^2} \tag{139}$$

The general solution of the variational system is then

$$x_1 = e^{(T_j/2)t}\left(c_{11}e^{i\omega t} + c_{12}e^{-i\omega t}\right)$$
$$x_2 = e^{(T_j/2)t}\left(c_{21}e^{i\omega t} + c_{22}e^{-i\omega t}\right) \qquad (140)$$

As T_j is negative the system is stable. Perturbed from its steady state it returns back to the same steady state through damped oscillations. The system is said to display a stable focus (Figure 23). For region III

$$T_j > 0, \qquad \Delta_j > 0, \qquad 4\Delta_j > T_j^2 \qquad (141)$$

and the two roots are complex with positive real parts. Perturbed from its steady state the system drifts towards a new steady state through amplified oscillations. The system has an unstable focus (Figure 23).

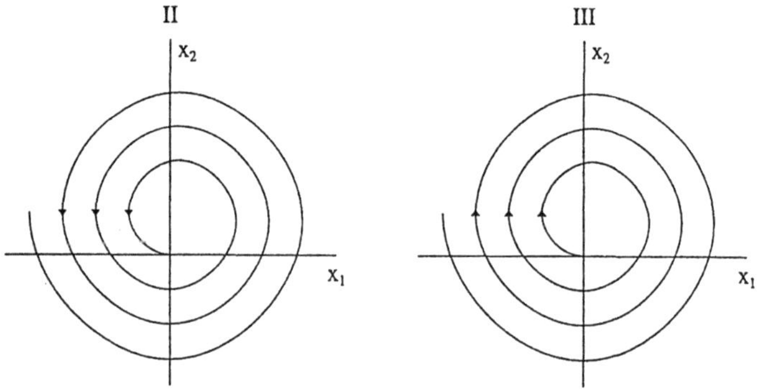

Fig. 23. Stable and unstable focus of a metabolic cycle

In regions V and VI one has

$$T_j > 0 \text{ (region V)}, \quad T_j < 0 \text{ (region VI)}, \quad \Delta_j < 0, \quad T_j^2 > 4\Delta_j \qquad (142)$$

The two roots are real and opposite in sign. The system is unstable and displays a saddle point (Figure 24).

A particularly important situation occurs if

$$T_j = 0, \qquad \Delta_j > 0, \qquad 4\Delta_j > T_j^2 \qquad (143)$$

Then the system displays sustained oscillations. The corresponding trajectory in the phase plane is an ellipse (Figure 24). This temporal organization of the system is called a "dissipative structure" by Prigogine.

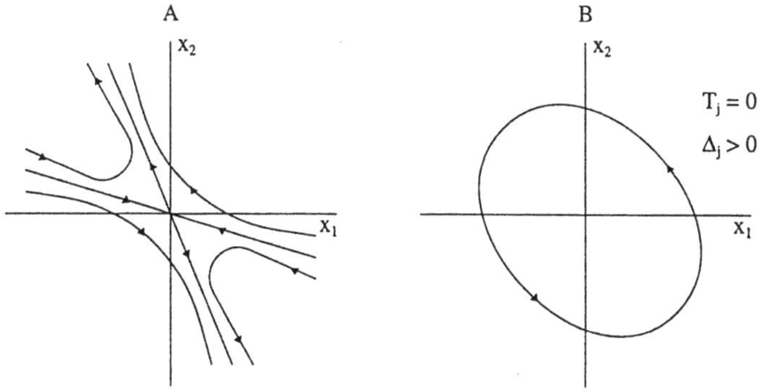

Fig. 24. Saddle point and limit cycle of a metabolic cycle

We are now in the position to determine what are the properties of the enzymes that are required to generate sustained oscillations. The trace and the determinant of the jacobian matrix of the system are

$$T_j = \frac{\partial f}{\partial S_1} + \frac{\partial g}{\partial S_2}$$
$$\Delta_j = \frac{\partial f}{\partial S_1}\frac{\partial g}{\partial S_2} - \frac{\partial g}{\partial S_1}\frac{\partial f}{\partial S_2} \tag{144}$$

If one assumes for simplicity that the two enzyme reactions are irreversible, the trace and the determinant have the following simple forms

$$T_j = -\left(\frac{\partial v_1}{\partial S_1} + \frac{\partial v_2}{\partial S_2} + k\right)$$
$$\Delta_j = k\frac{\partial v_1}{\partial S_1} \tag{145}$$

If, for instance, the two enzymes follow Michaelis-Menten kinetics

$$v_1 = \frac{S_{1\sigma}}{1 + S_{1\sigma}} \quad , \quad v_2 = \frac{S_{2\sigma}}{1 + S_{2\sigma}} \tag{146}$$

the two corresponding derivatives assume the forms

$$\frac{\partial v_1}{\partial S_1} = \frac{1}{(1 + S_{1\sigma})^2} \qquad \frac{\partial v_2}{\partial S_2} = \frac{1}{(1 + S_{2\sigma})^2} \tag{147}$$

As these two equations can take only positive values, the trace of the jacobian is always negative. Thus the enzyme system cannot display sustained oscillations. The simplest condition that may result in the appearance of these oscillations is that the enzyme E_2 be inhibited by an excess of substrate. This implies that

$$v_2 = \frac{S_{2\sigma}}{1 + S_{2\sigma} + \lambda\, S_{2\sigma}^2} \tag{148}$$

where λ is a constant. The corresponding derivative assumes the form

$$\frac{\partial v_2}{\partial S_2} = \frac{1 - \lambda S_{2\sigma}^2}{\left(1 + S_{2\sigma} + \lambda S_{2\sigma}\right)^2} \tag{149}$$

and may take positive or negative values. The trace of the jacobian may then be equal to zero and the system displays sustained oscillations. A number of biochemical systems are known to display these oscillations. Probably the one which has been most thoroughly investigated is the glycolytic pathway. These oscillations are periodic because the influx of matter in the system is constant (v_i = ct). However if this influx is itself a periodic function, the oscillations of the overall system may become aperiodic, or chaotic. Although the dynamics of the system is still expressed by deterministic differential equations, it is not possible to predict the state of the system at time $t + \Delta t$ if we know its state at time t. There exists now, in the biochemical and biological literature, a number of examples of this deterministic chaos, and it is very likely this number is going to increase in the years to come.

The general conclusions of this theoretical analysis are straightforward. The first conclusion is that these effects are to be expected only if the system is under non-equilibrium conditions. The spontaneous temporal organization of the system implies a dissipation of matter and energy (dissipative structure). The second conclusion is that this oscillatory dynamics can be expected to occur only if non-linear terms appear in the equations that describe the dynamic process. In the simple case considered above, this non-linear term is brought by the substrate inhibition of the second enzyme. The last conclusion is that the properties of the overall system are qualitatively novel with respect to those of any of the individual enzymes. For instance, whereas the overall system may follow oscillatory dynamics neither of the enzymes displays this type of behaviour and can be considered as oscillatory.

7.3. Dynamic organization of a metabolic cycle at the surface of a charged membrane

This type of reasoning may be easily extended to a metabolic cycle that occurs at the surface of a charged matrix, for instance a negatively charged membrane (Figure 25) (Kellershohn et al. 1991; Mulliert et al. 1991; Ricard et al. 1992). An enzyme (E_1) is located in the bulk phase whereas the other one (E_2) is located in the matrix. The influx and efflux of matter is supposed to take place outside the matrix. The two reaction intermediates S_1 and S_2, are partitioned between the matrix and the bulk phase. The expression of the electric partition coefficient is thus

$$\Pi = \frac{\sqrt{\Delta^2 + 4\,(S_{10}+S_{20})^2} + \Delta}{2(S_{10}+S_{20})} = \frac{2\,(S_{10}+S_{20})}{\sqrt{\Delta^2 + 4\,(S_{10}+S_{20})^2} - \Delta} \tag{150}$$

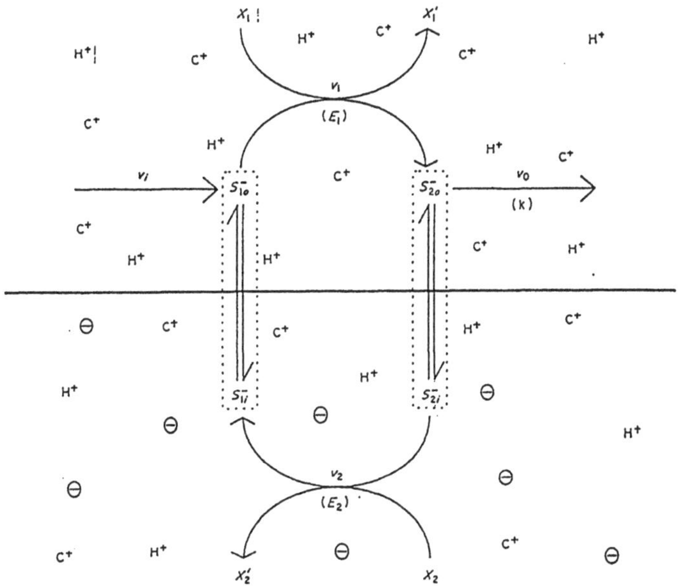

Fig. 25. A simple open metabolic cycle at the surface of a charged membrane (see text). The concentrations of X_1 and X_2 are assumed to be saturating as to cancel from the equations (from Ricard et al 1992)

One may also express the local concentrations of the metabolites S_1 and S_2 inside the matrix as a function of the corresponding bulk concentrations and of the fixed charge density. One finds

$$\begin{aligned} S_{1i} &= \frac{S_{10}}{2(S_{10}+S_{20})}\left\{\sqrt{\Delta^2 + 4\,(S_{10}+S_{20})^2} - \Delta\right\} \\ S_{2i} &= \frac{S_{20}}{2(S_{10}+S_{20})}\left\{\sqrt{\Delta^2 + 4\,(S_{10}+S_{20})^2} - \Delta\right\} \end{aligned} \tag{151}$$

If one assumes that the equilibration of S_1 and S_2 between the two phases is fast with respect to the other process, one has

$$\begin{aligned} \frac{d(S_{1o}+S_{1i})}{dt} &= \frac{dS_{1o}}{dt} + \frac{dS_{1i}}{dt} = v_i - v_1 + v_2 \\ \frac{d(S_{2o}+S_{2i})}{dt} &= \frac{dS_{2o}}{dt} + \frac{dS_{2i}}{dt} = v_1 - v_2 - v_o \end{aligned} \tag{152}$$

Moreover one may write

$$\begin{aligned} \frac{dS_{1i}}{dt} &= \left(\frac{\partial S_{1i}}{\partial S_{1o}}\right)\frac{dS_{1o}}{dt} + \left(\frac{\partial S_{1i}}{\partial S_{2o}}\right)\frac{dS_{2o}}{dt} \\ \frac{dS_{2i}}{dt} &= \left(\frac{\partial S_{2i}}{\partial S_{1o}}\right)\frac{dS_{1o}}{dt} + \left(\frac{\partial S_{2i}}{\partial S_{2o}}\right)\frac{dS_{2o}}{dt} \end{aligned} \tag{153}$$

and equation (152) may be rewritten as

$$\begin{aligned} \left\{1+\left(\frac{\partial S_{1i}}{\partial S_{1o}}\right)\right\}\frac{dS_{1o}}{dt} + \left(\frac{\partial S_{1i}}{\partial S_{2o}}\right)\frac{dS_{2o}}{dt} &= v_i - v_1 + v_2 \\ \left(\frac{\partial S_{2i}}{\partial S_{1o}}\right)\frac{dS_{1o}}{dt} + \left\{1+\left(\frac{\partial S_{2i}}{\partial S_{2o}}\right)\right\}\frac{dS_{2o}}{dt} &= v_1 - v_2 - v_o \end{aligned} \tag{154}$$

The four partial derivatives that appear in these equations may be considered as sensitivity coefficients, since they express how the local concentrations S_{1i} and S_{2i} vary when the corresponding bulk concentrations are varied. Differentiating, with respect to S_{1o} and S_{2o}, equations (151) allows one to derive the expression of these sensitivity coefficients. One finds

$$\begin{aligned} \sigma_{11} &= \frac{\partial S_{1i}}{\partial S_{1o}} = \frac{1}{\Pi}\left\{1+\frac{S_{1o}}{S_{1o}+S_{2o}}\,\frac{\Pi^2-1}{\Pi^2+1}\right\} \\ \sigma_{12} &= \frac{\partial S_{1i}}{\partial S_{2o}} = \frac{1}{\Pi}\,\frac{S_{1o}}{S_{1o}+S_{2o}}\,\frac{\Pi^2-1}{\Pi^2+1} \\ \sigma_{21} &= \frac{\partial S_{2i}}{\partial S_{1o}} = \frac{1}{\Pi}\,\frac{S_{2o}}{S_{1o}+S_{2o}}\,\frac{\Pi^2-1}{\Pi^2+1} \\ \sigma_{22} &= \frac{\partial S_{2i}}{\partial S_{2o}} = \frac{1}{\Pi}\left\{1+\frac{S_{2o}}{S_{1o}+S_{2o}}\,\frac{\Pi^2-1}{\Pi^2+1}\right\} \end{aligned} \tag{155}$$

These sensitivity coefficients are not all independent and must fulfil the following summation properties

$$\begin{aligned} \sigma_{11} + \sigma_{21} &= \sigma_{22} + \sigma_{12} = \frac{2\Pi}{\Pi^2 + 1} \\ \sigma_{11} + \sigma_{22} - \sigma_{12} - \sigma_{21} &= \frac{2}{\Pi} \end{aligned} \tag{156}$$

Defining dimensionless variables and parameters

$$\begin{array}{cccc} & \alpha_0 = \dfrac{S_{10}}{K_1} & \beta_0 = \dfrac{S_{20}}{K_2} & \\ \delta = \dfrac{\Delta}{K_2} & \varepsilon = \dfrac{K_1}{K_2} & \lambda = \dfrac{V_2}{V_1} & \theta = \dfrac{V_1 t}{K_1} \\ \nu_i = \dfrac{v_i}{V_1} & \nu_1 = \dfrac{v_1}{V_1} & \nu_2 = \dfrac{v_2}{V_1} & \nu_0 = \dfrac{v_0}{V_1} \end{array} \tag{157}$$

One may reexpress the four sensitivity coefficients in terms of these dimensionless quantities. One has

$$\begin{aligned} \sigma_{11} &= \frac{1}{\Pi}\left\{1 + \frac{\varepsilon\alpha_o}{\varepsilon\alpha_o + \beta_o}\,\frac{\Pi^2 - 1}{\Pi^2 + 1}\right\} \\ \sigma_{12} &= \frac{1}{\Pi}\,\frac{\varepsilon\alpha_o}{\varepsilon\alpha_o + \beta_o}\,\frac{\Pi^2 - 1}{\Pi^2 + 1} \\ \sigma_{21} &= \frac{1}{\Pi}\,\frac{\beta_o}{\varepsilon\alpha_o + \beta_o}\,\frac{\Pi^2 - 1}{\Pi^2 + 1} \\ \sigma_{22} &= \frac{1}{\Pi}\left\{1 + \frac{\beta_o}{\varepsilon\alpha_o + \beta_o}\,\frac{\Pi^2 - 1}{\Pi^2 + 1}\right\} \end{aligned} \tag{158}$$

The differential system (130) may thus be rewritten as

$$\frac{d}{d\theta}\begin{bmatrix}\alpha_o \\ \beta_o\end{bmatrix} = \frac{1}{\Omega}\begin{bmatrix} 1 + \sigma_{22} & -\sigma_{12} \\ -\sigma_{21}\varepsilon & (1 + \sigma_{11})\,\varepsilon \end{bmatrix}\begin{bmatrix} \nu_i - \nu_1 + \nu_2 \\ \nu_1 - \nu_2 + \nu_o \end{bmatrix} \tag{159}$$

where

$$\Omega = (1 + \sigma_{11})(1 + \sigma_{22}) - \sigma_{12}\sigma_{21} = \frac{(\Pi + 1)^3}{\Pi\,(\Pi^2 + 1)} \tag{160}$$

The normalized rates, ν_1, ν_2, ν_3 and ν_o, are indeed functions of α_o and β_o. Therefore the dynamic system may be rewritten as

$$\frac{d}{d\theta}\begin{bmatrix}\alpha_o \\ \beta_o\end{bmatrix} = \begin{bmatrix}F_1(\alpha_o, \beta_o) \\ F_2(\alpha_o, \beta_o)\end{bmatrix} \tag{161}$$

and a steady state is obtained if

$$\begin{bmatrix}F_1^*(\alpha_o, \beta_o) \\ F_2^*(\alpha_o, \beta_o)\end{bmatrix} = 0 \tag{162}$$

where the starred symbols refer to the steady state. As previously , stability analysis is based on the knowledge of the trace, T_j, the determinant , Δ_j of the jacobian matrix

$$J = \begin{bmatrix}\dfrac{\partial F_1^*}{\partial \alpha_o} & \dfrac{\partial F_1^*}{\partial \beta_o} \\[2ex] \dfrac{\partial F_2^*}{\partial \alpha_o} & \dfrac{\partial F_2^*}{\partial \beta_o}\end{bmatrix} \tag{163}$$

as well as on the discriminant $T_j^2 - 4\Delta_j$.

The number of steady states of the system depends upon the number of intersections of the curves representative of the two functions.

$$Y_1(\alpha_o) = \frac{\alpha_o}{1 + \alpha_o} \tag{164}$$

and

$$Y_2(\alpha_o) = \mu\beta_o^* + \frac{\lambda\beta_o^*}{\Pi + \beta_o^*} \tag{165}$$

where β_o^* is the reduced steady state concentration of S_2. Moreover Π may be expressed in terms of the reduced (or dimensionless) variables α_o and β_o^*. One has thus

$$\Pi = \frac{\sqrt{\delta^2 + 4\left(\varepsilon\alpha_o + \beta_o^*\right)} + \delta}{2\left(\varepsilon\alpha_o + \beta_o^*\right)} \tag{166}$$

As $\partial\Pi/\partial\alpha_o < 0$, $Y_2(\alpha_o)$ is of necessity an increasing function. Moreover the curves representative of the two functions $Y_1(\alpha_o)$ and $Y_2(\alpha_o)$ may intersect three times. This implies that the overall system may display three steady states when α_o^* is plotted *versus* β_o^* (Figure 26). Moreover two of these steady states are stable whereas the last one is unstable. As α_o^* (or the steady state of S_1) may take different values for the same value of β_o^* (or the steady state of S_2), the electric partition coefficient Π^*, under these steady state conditions, also displays a hysteresis (Figure 26). It is important at this stage to stress that both enzymes follow Michaelis-Menten kinetics. The remarkable property of multi-stability is therefore brought about by the electric repulsion effects exerted by the matrix on the mobile charges of S_1 and S_2.

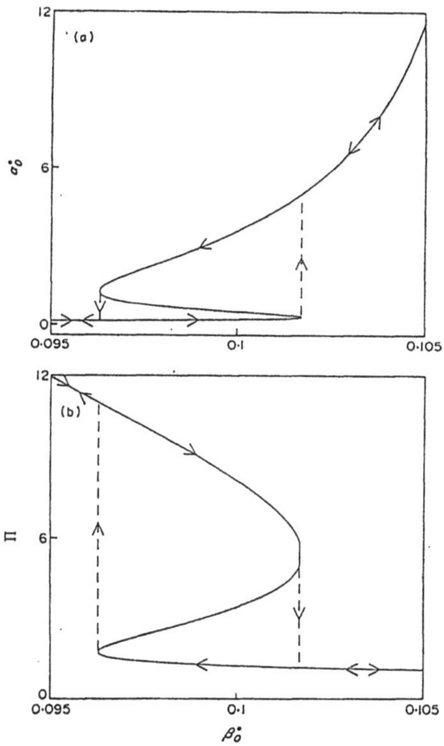

Fig. 26. Hysteresis of a substrate concentration and of the electric partition coefficient as a function of the other substrate concentration (see text). α_o and β_o represent the normalized substrate concentrations (from Ricard et al. 1992)

When the concentration of the substrate outside the matrix is increased the electric partition coefficient declines. This results in a change of proton concentration in the matrix and this change may in turn alter the enzyme activity. So far we have neglected this modulation of activity. But if this effect is taken into account it may give rise to surprising phenomena. Let us assume, for instance, that S_1 is a neutral molecule whereas S_2 is a monovalent anion. The electric partition coefficient then assumes the following expression

$$\Pi = \frac{\sqrt{\delta^2 + 4\beta_o^*} + \delta}{2\beta_o} = \frac{2\beta_o}{\sqrt{\delta^2 + 4\beta_o^*} - \delta} \tag{167}$$

Moreover one may define a dimensionless proton concentration γ_b as

$$\gamma_b = \frac{H_o}{K_b} \tag{168}$$

where K_b is the basic ionization constant of the enzyme-substrate complex. γ_b is thus a constant if the bulk proton concentration is held constant. If the apparent maximum reaction rate is sensitive to proton concentration and is increased when this concentration is increased according to a classical Dixon law, one has (Tipton and Dixon 1979; Dixon and Webb 1979)

$$\widetilde{V}_2 = \frac{V_2 H_i}{K_b + H_i} = \frac{V_2 H_o \Pi}{K_b + H_o \Pi} = \frac{V_2 \gamma_b \Pi}{1 + \gamma_b \Pi} \tag{169}$$

where $\widetilde{V}_2$ is the apparent maximum reaction rate. The two rate equations are

$$\begin{aligned} v_1(\alpha_o) &= \frac{\alpha_o}{1 + \alpha_o} \\ v_2(\beta_o) &= \lambda \frac{\gamma_b \Pi}{1 + \gamma_b \Pi} \frac{\beta_o}{\Pi + \beta_o} \end{aligned} \tag{170}$$

If $\gamma_b > 1$ the function $v_2(\beta_o)$ is monotonic and increasing , but if $\gamma_b < 1$ this function reaches at first a maximum and then decreases as β_o is increased. Thus although the enzyme follows by itself Michaelis-Menten kinetics, it may display apparent inhibition by excess substrate when bound to the matrix. This kinetic behaviour is due to the action of protons on the enzyme and its magnitude depends upon the fixed charge density of the matrix (Figure 27).

If, as previously assumed, S_1 is an uncharged molecule and S_2 a monoanion, one has

$$\begin{aligned} \sigma_{11} &= 1 \\ \sigma_{12} &= 0 \\ \sigma_{21} &= 0 \\ \sigma_{21} &= \frac{2\Pi}{\Pi^2 + 1} \end{aligned} \tag{171}$$

and the differential system (159) assumes the form

$$\frac{d}{d\theta}\begin{bmatrix} \alpha_o \\ \beta_o \end{bmatrix} = \frac{1}{\Omega}\begin{bmatrix} 1 + \sigma_{22} & 0 \\ 0 & 2\varepsilon \end{bmatrix}\begin{bmatrix} v_i - v_1 + v_2 \\ v_1 - v_2 + v_o \end{bmatrix} \tag{172}$$

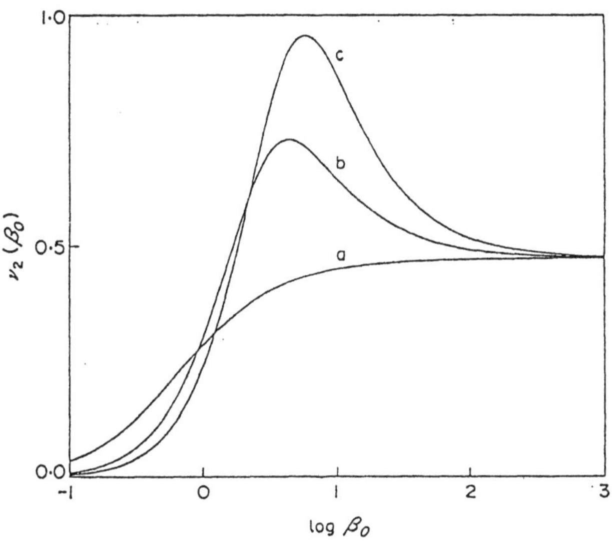

Fig. 27. Apparent inhibition of a bound enzyme through a change of proton concentration (see text). Curve a, b, and c pertain to different values of the dimensionless charge density: 1 , 10 and 20 respectively (from Ricard et al 1992)

where Ω is now equal to

$$\Omega = \frac{2\,(\Pi + 1)^2}{\Pi^2 + 1} \tag{173}$$

The steady state values, α_o^* and β_o^*, of the state variables, α_o and β_o, must be a solution of the following equations

$$\begin{aligned} v_i - v_1\,(\alpha_o) + v_2\,(\beta_o) &= 0 \\ v_1\,(\alpha_o) - v_2\,(\beta_o) - \mu\beta_o &= 0 \end{aligned} \tag{174}$$

which implies that

$$\beta_o^* = \frac{v_i}{\mu} \tag{175}$$

and that α_o^* be a solution of the equation

$$\frac{\alpha_o}{1 + \alpha_o} = \mu\beta_o^* + \lambda\,\frac{\gamma_b\Pi^*}{1 + \gamma_b\Pi^*}\;\frac{\beta_o^*}{\Pi^* + \beta_o^*} \tag{176}$$

Stability analysis of the above dynamic system (172) allows one to express the trace, the determinant of the jacobian, as well as the expression $T_j^2 - 4\Delta_j$. One finds

$$T_j = -\left\{\frac{1}{2}\frac{\partial v_1^*}{\partial \alpha_o^*} + \frac{\varepsilon\,(\Pi^{*2}+1)}{(\Pi^*+1)^2}\left(\frac{\partial v_2^*}{\partial \beta_o^*} + \mu\right)\right\}$$
$$\Delta_j = \frac{\varepsilon\,(\Pi^{*2}+1)}{2\,(\Pi^{*2}+1)^2}\frac{\partial v_1^*}{\partial \alpha_o^*}\,\mu \qquad (177)$$

It thus appears that the determinant Δ_j is always positive and that the trace T_j may be either positive or negative. A necessary but, by no means, sufficient condition in order to have this trace positive is that $\partial v_2^* / \partial \beta_o^* < 0$. This implies that the function v_2^* be inhibited by an excess of substrate. This is precisely the situation described in Figure 28. If the trace is positive the system may display oscillations of the concentrations of the two metabolites S_1 and S_2 (or α_o and β_o), as well as of the electric partition coefficient and of the sensitivity coefficient (Figure 28).

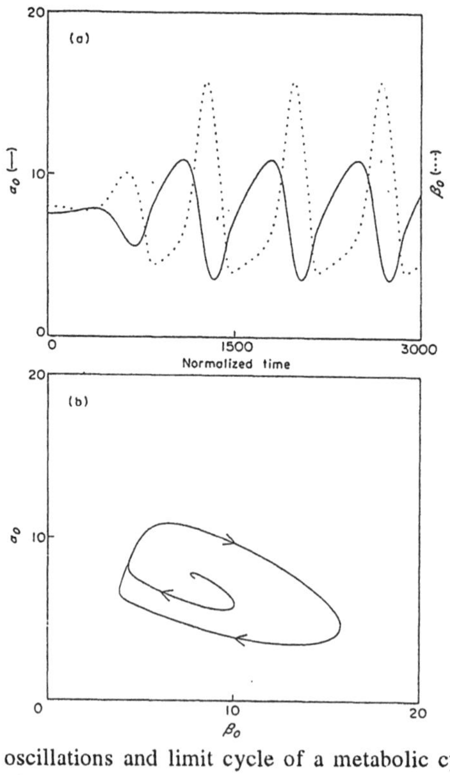

Fig. 28. Periodic oscillations and limit cycle of a metabolic cycle at the surface of a membrane. **a** Sustained oscillations of the reaction intermediates. **b** Limit cycle (from Ricard et al. 1992)

In one of the previous sections hysteretic effects have been described as arising from the coupling between diffusion and an enzyme reaction inhibited by an excess of substrate. The non-linear term that is required to generate hysteresis originates from the intrinsic kinetic properties of the enzyme that is inhibited by excess substrate. In the present case of enzymes occurring at the surface of a charged membrane, the origin of hysteresis is not to be found in the intrinsic properties of any of the enzymes, that both follow Michaelis-Menten kinetics, but rather in the interplay between electric repulsion of the substrate by the matrix and the simple kinetic properties of the enzymes.

The existence of hysteresis of the substrate concentration at the surface of a membrane brings about a hysteresis of the electric partition coefficient when the input of matter in the system is varied. This implies that the electric repulsion effect exerted by the surface of the membrane for the same input of matter is different depending on whether this input value has been reached after an increase or a decrease. This implies that the surface of the membrane may be viewed as a biosensor.

Exactly as for hysteresis, sustained oscillations require dynamic equations that contain non-linear terms. In the previous section it has been shown that these non-linear terms were introduced in the system through the intrinsic properties of one of the enzymes which is inhibited by excess substrate. The present situation is equivalent to an inhibition by excess of the substrate for, as the substrate concentration is increased, the electric partition coefficient changes as well as the local proton concentration and this may result in an inhibition of enzyme activity. Therefore, as for hysteresis, the non-linear terms required for oscillations originate from the interplay between the intrinsic properties of one of the enzymes and of the electric repulsion exerted by the charged substrate.

8. Control of multi-enzyme complexes

It was implicitly assumed so far that two enzymes catalysing distinct reactions were physically separate molecular entities. But it may perfectly well occur that these two enzymes be physically associated as a bi-enzyme complex. It often occurs that these multi-enzyme complexes catalyse consecutive reactions of the same metabolic pathway. It is therefore of interest to know whether the physical association of two enzymes that catalyse different, for instance consecutive, reactions may alter the overall kinetics of the multi-molecular aggregate (Barnes and Weitzman 1986; Beudeker and Kuener 1981; Friedrich 1985; Gontero et al. 1988; Keleti et al. 1977; Keleti 1984; Kirschner and Bisswanger 1976; Kurganov et al. 1985; Mowbray and Moses 1976; Nicholson et al. 1986, 1987; Persson and Johansson 1989; Reed et al. 1975; Reed 1981; Robinson and Srere 1985; Robinson et al. 1987; Sainis and Harris 1986; Sainis et al. 1989; Salerno et al.

1982; Shimakata and Stumpf 1982; Srere 1967; Srere 1972; Srere 1985 a and b; Srere 1987; Srivastava and Bernhard 1987; Tompa et al. 1987; Volpe and Vagelos 1976; Walsh et al. 1977; Welch 1977; Wieland et al. 1979; Wakil et al. 1989). It has often been speculated, and in some cases experimentally demonstrated, that channelling of a reaction intermediate from one enzyme to another one may occur. This interesting phenomenon may take place through different molecular mechanisms: diffusion of the reaction intermediate at the surface of the enzyme, conformation change of the enzyme complex, direct transfer of the intermediate from one site to the other thanks to a coenzyme of one of the enzymes (Hammes 1981).

Even when channelling does not occur, the simple association of two different enzymes may result in significant alterations of the kinetics of these two enzymes. A phenomenological model which illustrates this situation is shown in Figure 29. The relevant steady state equation for the appearance of any of the two products, P_1 and P_2, involves squared terms in both S_1 and S_2 if no assumption is made as to the nature of energy interactions exerted between enzymes.

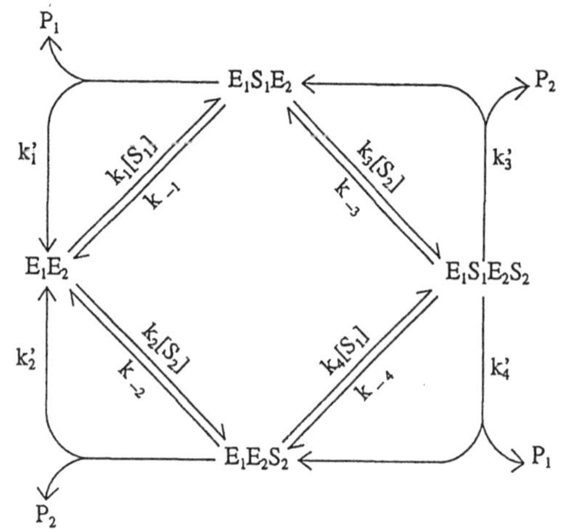

Fig. 29. Kinetic process carried out by a bi-enzyme complex (see text). In all generality this kinetic scheme should generate plots that do not fit Michaelis-Menten kinetics

8.1. Generalized microscopic reversibility and multi-enzyme complexes

There are simple thermodynamic conditions however which result in a dramatic simplification of the reaction rate expression. These conditions, are known under the term of generalized microscopic reversibility (Whitehead 1976). Thus, in the case of a two-enzyme complex the reaction rate follows hyperbolic kinetics with respect to the appearance of the two products if

$$\frac{k'_2}{k_{-2}} = \frac{k'_3}{k_{-3}} \quad \text{and} \quad \frac{k'_1}{k_{-1}} = \frac{k'_4}{k_{-4}} \tag{178}$$

If these equalities were purely fortuitous it would have been of little interest to discuss them. It is only because relations (178) may have a thermodynamic significance that it is worth discussing these conditions of simplification.

Let us consider a multi-enzyme complex, for instance a two-enzyme complex, the free energy of activation, $\Delta G^{\neq}$, associated with a rate constant, k, may be written as

$$\Delta G^{\neq} = \Delta G^{\neq *} + U_\gamma - U_\tau \tag{179}$$

where $\Delta G^{\neq *}$ is the so-called intrinsic free energy of activation, that is what this free energy would be if the corresponding enzyme were naked. U_γ is an energy contribution which expresses how enzyme-enzyme interactions stabilize (U_γ positive) or destabilize (U_γ negative), in the ground state, the two-enzyme complex with respect to the isolated enzymes. U_τ is also an energy contribution which represents how enzyme interaction within the complex stabilize (U_τ positive) or destabilize (U_τ negative) the transition state relative to that of the isolated enzymes. In figure 30 is illustrated the part played by these contributions U_γ and U_τ, in the expression of the free energy of activation.

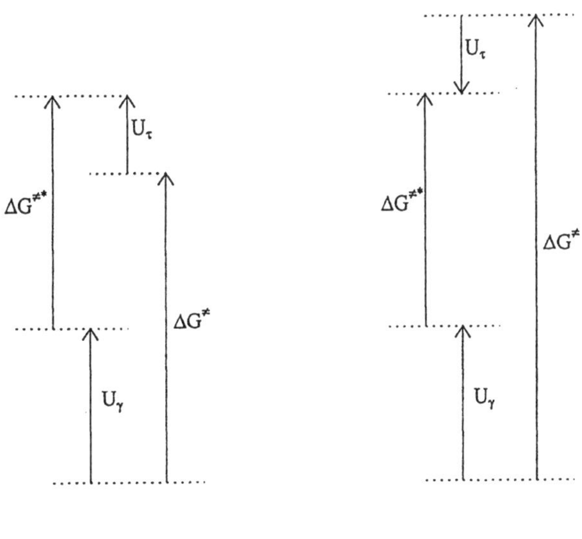

Fig. 30. Effect of enzyme-enzyme interaction on the stabilization or the destabilization of a reaction process (see text). **A** Enzyme-enzyme interaction stabilizes the ground and the transition states (U_γ and U_τ positive). **B** Enzyme-enzyme interaction stabilizes the ground state and destabilizes the transition state (U_γ positive, U_τ negative)

The rate constant associated with this energy of activation is thus

$$k = k^* \exp\{-(U_\gamma - U_\tau)/RT\} \tag{180}$$

where k^* is the intrinsic rate constant of the corresponding process carried out by the isolated enzyme. The condition that generates generalized microscopic reversibility is thus that the energy contribution U_γ and U_τ be the same for substrate release and for catalysis (Figure 30). This, in turn, implies that the enzyme conformation induced by the binding of two transition states $S^{\neq}$ and $X^{\neq}$ is approximately the same.

8.2. The stoichiometry of polypeptide chains in multi-enzyme complexes

Recent studies have shown that when a polymeric enzyme is made up of different polypeptide chains, the stoichiometry of these polypeptide chains may be different depending on whether the enzyme is naked or associated with other enzymes as a multi-enzyme complex. This appears to be the case of ribulose bisphosphate carboxylase-oxygenase. This enzyme is in part free in the stroma of the chloroplast, and in part associated with at least four other enzymes, namely phosphoribulokinase, glyceraldehyde phosphate dehydrogenase, phosphoribose isomerase, phosphoglycerate kinase (Gontero et al. 1988). This complex may be purified or detected in the supernatant of osmotically shocked chloroplasts by antibodies specifically raised against either phosphoribulokinase, glyceraldehyde phosphate dehydrogenase or ribulose bisphosphate carboxylase-oxygenase. Naked glyceraldehyde phosphate dehydrogenase is a tetramer made up of two different types of polypeptide chains and has a structure of the A_2B_2 type. Free phosphoribulokinase is a dimer made up of identical subunits and ribulose bisphosphate carboxylase-oxygenase has a more complex quaternary structure with eight large and eight small subunits. The combined use of immunochemical and densitometric techniques shows that within the complex one has two subunits of phosphoribulokinase, two A and two B subunits of glyceraldehyde phosphate dehydrogenase, two large and four small subunits of ribulose bisphosphate carboxylase-oxygenase. Of particular interest, in this molecular edifice, is thus the stoichiometry and the number of the polypeptide chains of ribulose bisphosphate carboxylase-oxygenase which are completely different from those existing in the naked enzyme. This enzyme follows Michaelis-Menten kinetics either when it is free, or embedded in the multi-complex, but in the latter case the Vmax is considerably increased and the Km decreased with respect to what is obtained with the free enzyme. It is therefore important to offer a rationale for this dramatic change of activity that may be observed upon insertion of an enzyme into a complex. This rationale may be found in what is now called structural kinetics.

8.3. The principles of structural kinetics of oligomeric enzymes and of multi-enzyme complexes

Recently, an attempt has been made to understand how information transfer between subunits (or polypeptide chains) of a polymeric enzyme molecule may alter kinetic constants and therefore the enzyme reaction rate. This attempt has been termed structural kinetics (Ricard 1985; Ricard and Cornish-Bowden 1987; Ricard and Noat 1985; 1986 b; Ricard et al. 1990; Giudici-Orticoni et al. 1990 a and b). Moreover the basic idea of structural kinetics may be extended to the case where the oligomeric enzyme which is involved in a catalytic process is associated with another enzyme that does not catalyse any chemical reaction, for the corresponding substrates are lacking. In the following, for simple convenience, the oligomeric enzyme that catalyses the chemical process is called the "active" enzyme whereas the other enzyme of the complex is defined as "inactive". As information transfer between polypeptide chains occurs upon substrate binding, substrate release, or catalysis there is no information transfer between the "inactive" and the "active" enzyme within the multimolecular complex. This "inactive" enzyme, however, exerts an effect on the rate of the conformational transition associated with catalysis. This "inactive" enzyme thus modulates the overall catalytic activity within the complex. The free energy of activation, $\Delta G^{\neq}$, associated with a given reaction is thus modulated by the information transfer that occurs between identical subunits of the oligomeric enzyme in the complex and by an alteration of the rate of the conformational transitions of these subunits exerted through the interactions that exist between the "active" subunits and the catalytically "inactive" enzyme molecule(s). One may still express a rate process conditioned by a polymeric enzyme, or by an enzyme complex, by equation (179) but $\Delta G^{\neq *}$ has a meaning different from the one used in equation (179). It represents what the free energy of activation would be if the subunits of the enzyme were independent. U_γ and U_τ have also a significance which is slightly different from the one defined previously. U_γ is an energy contribution which expresses how subunit interactions stabilize (U_γ positive) or destabilize (U_γ negative), in the ground state, the oligomeric enzyme with respect to the isolated subunit. U_τ is also an energy contribution which represents how subunit interactions, stabilize (U_τ positive) or destabilize (U_τ negative) the transition state of the enzyme relative to that of the isolated subunit.

Within the frame of these definitions the energies of stabilization, of destabilization, U_γ or U_τ, may result from three types of effects. The first one, $\Sigma({}^{\alpha}\Delta G)$, is exerted by the association of "active" subunits on the rate of the conformational transition of these subunits, which carries the chemical reaction. This energy is equivalent to the free energy change that would accompany the dissociation of the oligomeric enzyme into the naked "active" subunits, on assuming that no conformation change of the subunits occurs during the dissociation and that the "inactive" enzyme does not affect this dissociation process. The second type of energy contribution, $\Sigma({}^{\sigma}\Delta G)$, is exerted through mutual distortion of the subunits (intersubunit strain). The last energy contribution $\Sigma({}^{\mu}\Delta G)$, is the one exerted by the "inactive" enzyme(s) on the rate of the conformational transition of the "active" subunits which carry the chemical reaction. One has thus

$$U_{\gamma} = \sum ({}^{\alpha}\Delta G) + \sum ({}^{\sigma}\Delta G) + \sum ({}^{\mu}\Delta G) \tag{181}$$

and a similar relation holds for U_{τ}.

If the oligomeric "active" enzyme, in the multimolecular complex, is made up of n interacting subunits that exist in two unstrained conformation states, A and B, if there exist ℓ subunits in conformation B and therefore n-ℓ subunits in conformation A, the energy contribution, $\Sigma({}^{\alpha}\Delta G)$, assumes the form

$$\sum ({}^{\alpha}\Delta G) = \binom{n-\ell}{2}({}^{\alpha}\Delta G_{AA}) + \binom{\ell}{2}({}^{\alpha}\Delta G_{BB}) + \ell\,(n-\ell)\,({}^{\alpha}\Delta G_{AB}) \tag{182}$$

where $({}^{\alpha}\Delta G_{AA})$, $({}^{\alpha}\Delta G_{BB})$ and $({}^{\alpha}\Delta G_{AB})$ represent the energies required to dissociate unstrained dimers AA, BB and AB into their subunits.

These energies, are thus ideal for it is assumed that no quaternary constraint exists between these subunits and that the rate of their conformational transition is not affected by their interactions with the "inactive" enzyme(s).

The energy of inter-subunit strain, $\Sigma({}^{\sigma}\Delta G)$, involves two types of contributions, $({}^{\beta}\Delta G)$, which represents the energy required to break weak bonds that maintain the subunits in a strained state, $({}^{\sigma}\Delta G_{A})$ and $({}^{\sigma}\Delta G_{B})$ which are the energies required to strain relieved A and B subunits to the state they have in the strained enzyme. As for the contribution $\Sigma({}^{\alpha}\Delta G)$, it is not taken into account the effect of the "inactive" subunits. The expression of this new energy contribution is then

$$\begin{aligned}\sum ({}^{\sigma}\Delta G) = {} & \binom{n-\ell}{2}({}^{\beta}\Delta G_{AA}) + \binom{\ell}{2}({}^{\beta}\Delta G_{BB}) + \ell\,(n-\ell)\,({}^{\beta}\Delta G_{AB}) \\ & -(n-\ell)\,({}^{\sigma}\Delta G_{A}) - \ell\,({}^{\sigma}\Delta G_{B})\end{aligned} \tag{183}$$

The third energy contribution, $\Sigma({}^{\mu}\Delta G)$, expresses how the "inactive" enzyme in the complex modulates the rate of the conformational transition of the strained active subunits, A' and B', without taking account of the effects of quaternary constraints between identical "active" subunits, as well as that of subunit interactions on the dynamics of conformational transitions. If the "inactive" enzyme is in interaction with p "active" subunits, the expression of this energy contribution assumes the form

$$\sum ({}^{\mu}\Delta G) = p\,(n - \ell)\,({}^{\mu}\Delta G_{A'}) + p\,\ell\,({}^{\mu}\Delta G_{B'}) \qquad (184)$$

where $({}^{\mu}\Delta G_{A'})$ and $({}^{\mu}\Delta G_{B'})$ represent the free energy of dissociation of the "inactive" enzyme from the strained conformations A' and B' of two "active" subunits. It then becomes possible to express how subunit interactions within an oligomeric enzyme, or a multi-enzyme complex, alter the value of the catalytic rate constants, k_i', or that of the reciprocal of the Michaelis constants, $\overline{K}_i$. Without any assumption, however, this expression is extremely complex and untractable in practice. Therefore structural kinetics, whether applied to oligomeric enzymes, or to enzyme complexes, relies upon a number of postulates that are not gratuitous assumptions, or *ad hoc* hypotheses, but rather simplifying assumptions that are physically sound. Structural kinetics of enzyme complexes is then based on five postulates:

First postulate: Inter-subunit strain is relieved upon reaching the transition states of the reaction. This postulate is an extrapolation to polymeric enzymes, or to enzyme complexes, of what is known to occur for monomeric enzymes. As a matter of fact it is now considered that intra-molecular strain is relieved in the transition states (Wolfenden 1969, Jencks 1975). This represents the very basis of catalysis. It is thus obvious that for this relief to occur, inter-subunit strain has to be abolished as well.

Second postulate: In the absence of inter-subunit strain, the subunits assume either of only two conformations called A and B. This postulate of the minimum number of conformations is exactly the one already made by Monod et al. (1965) as well as by Koshland et al. (1966). B represents the subunit which has bound the substrate, and A the one which is unliganded.

Third postulate: The subunit which has bound a transition state has a conformation, A, identical to that of the unliganded subunit. This is in agreement with the view that the unstrained active site of an enzyme is "complementary" to the transition state of the reaction (Jencks 1975), and with the "economy" principle (Occam's razor) considered above.

Fourth postulate: In a constrained oligomeric enzyme, or a multi-enzyme complex, the potential energy generated by the intersubunit strain is evenly stored in all the subunits, whether their subunits have bound a ligand or not. This is again an illustration of the same "economy" principle considered above.

Fifth postulate: Within a multi-enzyme complex the interactions of two different enzymes are rather weak in such a way that the "inactive" enzyme does not contribute to the strain of the oligomeric "active" enzyme. The effect of this "inactive" enzyme is thus solely to alter the rate of the conformational transitions of the "active" subunits which carry out catalysis. This postulate, which is again an "economy" principle, implies that the quaternary constraints between "active" subunits are the same whatever the oligomeric enzyme is naked or associated with other polypeptide chains. As shown in Figure 31 this is equivalent to assuming that the "inactive" enzyme has the same affinity for the strained and for the unstrained "active" subunits.

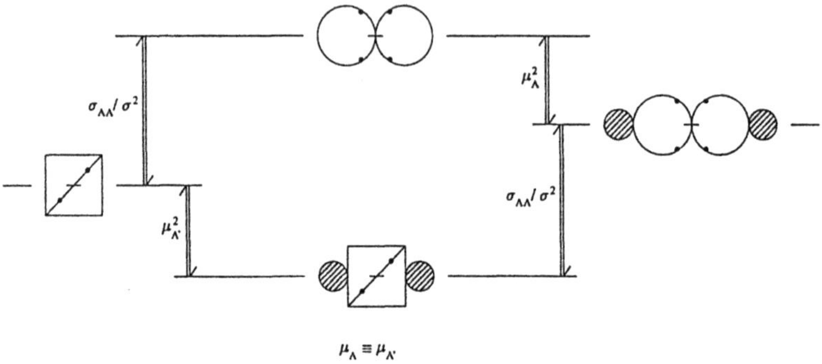

Fig. 31. Thermodynamic "box" that demonstrates that an "inactive" enzyme which does not contribute to the strain of the active subunits binds the strained and the unstrained "active" subunits equally well. The dashed circles represent the "inactive" enzyme. The "active" subunits exist in a strained (triangle) and in an unstrained (circle) state. The strained state is maintained through weak bonds between subunits (dots)

If an enzyme form, E_{i-1}, is converted upon binding of a substrate molecule into another form, E_i, able to carry out catalysis, three types of rate processes should be considered: the binding of the substrate (rate constant k_i), the release of that substrate (rate constant k_{-i}), the catalysis (rate constant k_i'). The postulates above allow one to express the reciprocal of the Michaelis constant associated with this process as

$$\overline{K}_i = \overline{K}^* \exp\{-(U_{i-1,\gamma} - U_{i,\gamma})/RT\} \tag{185}$$

Similarly the rate constant of catalysis, k_i', is expressed as

$$k_i' = k_c^* \exp\{-(U_{i,\gamma} - U_{i-1,\tau})/RT\} \tag{186}$$

In expression (185) $\overline{K}^*$ is the reciprocal of intrinsic Michaelis constant, that is what this constant would be if the enzymes were monomeric. As defined previously k_c^* is the intrinsic catalytic constant, that is what this rate constant would be if no subunit, or protomer, interaction were occurring. The application of the postulates of structural kinetics allows one to write

$$U_{i-1,\gamma} - U_{i,\gamma} = (n-i)\left\{\left({}^{\alpha}\Delta G_{AA}\right) + \left({}^{\beta}\Delta G_{AA}\right)\right\} - (i-1)\left\{\left({}^{\alpha}\Delta G_{BB}\right) + \left({}^{\beta}\Delta G_{BB}\right)\right\}$$
$$- (n-2i+1)\left({}^{\alpha}\Delta G_{AB}\right) + p\left({}^{\mu}\Delta G_{A}\right) - p\left({}^{\mu}\Delta G_{B}\right) \tag{187}$$

and similarly

$$U_{i,\gamma} - U_{i-1,\tau} = -(n-i)\left({}^{\alpha}\Delta G_{AA}\right) + (i-1)\left({}^{\alpha}\Delta G_{BB}\right) + (n-2i+1)\left({}^{\alpha}\Delta G_{AB}\right)$$
$$+\binom{n-i}{2}\left({}^{\beta}\Delta G_{AA}\right) + \binom{i}{2}\left({}^{\beta}\Delta G_{BB}\right) + i(n-i)\left({}^{\beta}\Delta G_{AB}\right)$$
$$- n\left({}^{\sigma}\Delta G\right) - p\left({}^{\mu}\Delta G_{A}\right) + p\left({}^{\mu}\Delta G_{B}\right) \tag{188}$$

The terms $({}^{\sigma}\Delta G)$, $({}^{\mu}\Delta G_A)$ and $({}^{\mu}\Delta G_B)$ appear as a consequence of the fourth and fifth postulates above for one must have

$$\begin{aligned} \left({}^{\sigma}\Delta G_{A}\right) &= \left({}^{\sigma}\Delta G_{B}\right) = \left({}^{\sigma}\Delta G\right) \\ \left({}^{\mu}\Delta G_{A'}\right) &= \left({}^{\mu}\Delta G_{A}\right) \\ \left({}^{\mu}\Delta G_{B'}\right) &= \left({}^{\mu}\Delta G_{B}\right) \end{aligned} \tag{189}$$

One may thus define the following dimensionless coefficients as

$$\begin{aligned} \alpha_{AA} &= \exp\left\{-\left({}^{\alpha}\Delta G_{AA}\right)/RT\right\} \\ \alpha_{BB} &= \exp\left\{-\left({}^{\alpha}\Delta G_{BB}\right)/RT\right\} \\ \alpha_{AB} &= \exp\left\{-\left({}^{\alpha}\Delta G_{AB}\right)/RT\right\} \\ \sigma_{AA} &= \exp\left\{-\left({}^{\beta}\Delta G_{AA}\right)/RT\right\} \\ \sigma_{BB} &= \exp\left\{-\left({}^{\beta}\Delta G_{BB}\right)/RT\right\} \\ \sigma_{AB} &= \exp\left\{-\left({}^{\beta}\Delta G_{AB}\right)/RT\right\} \\ \sigma &= \exp\left\{-\left({}^{\sigma}\Delta G\right)/RT\right\} \\ \mu_{A} &= \exp\left\{-\left({}^{\mu}\Delta G_{A}\right)/RT\right\} \\ \mu_{B} &= \exp\left\{-\left({}^{\mu}\Delta G_{B}\right)/RT\right\} \end{aligned} \tag{190}$$

and the expression of the reciprocal of the Michaelis constant, $\overline{K}_i$, assumes the form

$$\overline{K}_i = \overline{K}^* \frac{\alpha_{AA}^{n-i}}{\alpha_{BB}^{i-1}\,\alpha_{AB}^{n-2i+1}} \frac{\sigma_{AA}^{n-i}}{\sigma_{BB}^{i-1}\,\sigma_{AB}^{n-2i+1}} \frac{\mu_A^p}{\mu_B^p} \tag{191}$$

Similarly the catalytic rate constant, k_i', may be written as

$$k_i' = k_c^* \frac{\alpha_{BB}^{i-1}\,\alpha_{AB}^{n-2i+1}}{\alpha_{AA}^{n-i}} \frac{\sigma_{AA}^{(n-i)(n-i-1)/2}\,\sigma_{BB}^{i(i-1)/2}\,\sigma_{AB}^{i(n-i)}}{\sigma^n} \frac{\mu_B^p}{\mu_A^p} \tag{192}$$

Equations (191) and (192) are thus structural for they express how Michaelis and rate constants depend upon α, σ and μ thermodynamic parameters which represent how information transfer between subunits and interaction with foreign polypeptide chains alters binding and catalytic properties of an enzyme. In the case, for instance, of a tetramer made up of identical subunits that are all in interaction, the structural reaction rate is

$$v = 4\,k_c^*\,[E]_o \frac{\sigma_{AA}^6}{\sigma^4}\,\overline{K}^*\,[S] \frac{\overline{\Pi}_\alpha(3)}{\overline{\Pi}_{\alpha,\sigma}(4)} \tag{193}$$

where $[E]_o$ is the total enzyme concentration, $\overline{\Pi}_\alpha(3)$ and $\overline{\Pi}_{\alpha,\sigma}(4)$ polynomials which may be written as

$$\overline{\Pi}_{\alpha,\sigma}(4) = 1 + 4\frac{\alpha_{AA}^3}{\alpha_{AB}^3}\frac{\sigma_{AA}^3}{\sigma_{AB}^3}\frac{\mu_A^p}{\mu_B^p}\overline{K}^*[S] + 6\frac{\alpha_{AA}^5}{\alpha_{AB}^4\,\alpha_{BB}}\frac{\sigma_{AA}^5}{\sigma_{AB}^4\,\sigma_{BB}}\frac{\mu_A^{2p}}{\mu_B^{2p}}\overline{K}^{*2}[S]^2$$
$$+ 4\frac{\alpha_{AA}^6}{\alpha_{AB}^3\,\alpha_{BB}^3}\frac{\sigma_{AA}^6}{\sigma_{AB}^3\,\sigma_{BB}^3}\frac{\mu_A^{3p}}{\mu_B^{3p}}\overline{K}^{*3}[S]^3 + \frac{\alpha_{AA}^6}{\alpha_{BB}^6}\frac{\sigma_{AA}^6}{\sigma_{BB}^6}\frac{\mu_A^{4p}}{\mu_B^{4p}}\overline{K}^{*4}[S]^4 \tag{194}$$

and

$$\overline{\Pi}_\alpha(3) = 1 + 3\frac{\alpha_{AA}^3}{\alpha_{AB}^3}\frac{\mu_A^p}{\mu_B^p}\overline{K}^*[S] + 3\frac{\alpha_{AA}^5}{\alpha_{AB}^4\,\alpha_{BB}}\frac{\mu_A^{2p}}{\mu_B^{2p}}\overline{K}^{*2}[S]^2$$
$$+ \frac{\alpha_{AA}^6}{\alpha_{AB}^3\,\alpha_{BB}^3}\frac{\mu_A^{3p}}{\mu_B^{3p}}\overline{K}^{*3}[S]^3 \tag{195}$$

Expression (194) represents how the population of enzyme molecules is distributed, during the steady state of the reaction, over the allowed energy levels of the ground state. Similarly the polynomial (195) expresses how these enzyme molecules are distributed among the energy levels of the transition state.

8.4. Thermodynamics of the alteration of the kinetic behaviour of an oligomeric enzyme within a multi-enzyme complex

Results that have been mentioned above show that ribulose bisphosphate carboxylase-oxygenase within a five-enzyme complex displays a kinetic behaviour which is dramatically altered with respect to the one it has in the free state. Neither in this free state, nor associated with other proteins does this enzyme display any co-operativity, but its Vm is increased and its Km decreased when associated with the other enzymes. The aim of this section is to offer a rationale for this type of effect.

As an illustration of the influence exerted through the interactions of polypeptide chains belonging to different enzymes, one may consider, for instance, an enzyme which is tetrameric in the free state but which becomes dimeric when associated with another protein in a complex. Whatever its state, the enzyme displays no co-operativity. Moreover the Vm is enhanced and the Km decreased within the multi-enzyme complex.

In order to explain the lack of co-operativity, the polynomials $\overline{\Pi}_{\alpha,\sigma}$ and $\overline{\Pi}_{\alpha}$ for both the naked enzyme and the complex should be factorized so as to keep only the linear S terms in the reaction rate equation. There are two obvious conditions that allow this factorization, either

$$\begin{aligned} \alpha_{AA} &= \alpha_{AB} = \alpha_{BB} \\ \sigma_{AA} &= \sigma_{AB} = \sigma_{BB} \end{aligned} \tag{196}$$

or

$$\begin{aligned} \alpha_{AB}^{2} &= \alpha_{AA}\,\alpha_{BB} \\ \sigma_{AA} &= \sigma_{AB} = \sigma_{BB} \end{aligned} \tag{197}$$

The conditions (196) are not acceptable for they imply that the enzyme should have the same Km, or the same apparent affinity constant $\overline{K}_s$, in either states. If conditions (197) apply then

$$\begin{aligned} \overline{\Pi}_{\alpha,\sigma}(4) &= \left(1 + \frac{\alpha_{AA}^{3}}{\alpha_{AB}^{3}}\,\overline{K}^{*}\,[S]\right)^{4} \\ \overline{\Pi}_{\alpha}(3) &= \left(1 + \frac{\alpha_{AA}^{3}}{\alpha_{AB}^{3}}\,\overline{K}^{*}\,[S]\right)^{3} \end{aligned} \tag{198}$$

for the naked tetrameric enzyme and

$$\begin{aligned} \overline{\Pi}'_{\alpha,\sigma}(2) &= \left(1 + \frac{\alpha_{AA}}{\alpha_{AB}}\,\frac{\mu_{A}^{2}}{\mu_{B}^{2}}\,\overline{K}^{*}\,[S]\right)^{2} \\ \overline{\Pi}'_{\alpha}(1) &= 1 + \frac{\alpha_{AA}}{\alpha_{AB}}\,\frac{\mu_{A}^{2}}{\mu_{B}^{2}}\,\overline{K}^{*}\,[S] \end{aligned} \tag{199}$$

for the complex. Therefore the structural reaction rate for the naked enzyme is

$$v = 4\,\frac{\sigma_{AA}^{6}}{\sigma^{4}}\;\frac{k_c^{*}[E]_o\,\overline{K}^{*}[S]}{1+\dfrac{\alpha_{AA}^{3}}{\alpha_{AB}^{3}}\overline{K}^{*}[S]} \tag{200}$$

and the corresponding reaction rate for the complex is

$$v' = 2\,\frac{\sigma_{AA}}{\sigma^{2}}\;\frac{k_c^{*}[E]_o\,\overline{K}^{*}[S]}{1+\dfrac{\alpha_{AA}}{\alpha_{AB}}\dfrac{\mu_A^{2}}{\mu_B^{2}}\overline{K}^{*}[S]} \tag{201}$$

The Michaelis constant of the free enzyme is thus

$$K_m = \frac{\alpha_{AB}^{3}}{\alpha_{AA}^{3}}\,\frac{1}{\overline{K}^{*}} \tag{202}$$

and the corresponding constant for the complex assumes the form

$$K_m' = \frac{\alpha_{AB}}{\alpha_{AA}}\,\frac{\mu_B^{2}}{\mu_A^{2}}\,\frac{1}{\overline{K}^{*}} \tag{203}$$

Thus the condition Km > Km' is equivalent to

$$\frac{\alpha_{AB}^{2}}{\alpha_{AA}^{2}} > \frac{\mu_B^{2}}{\mu_A^{2}} \tag{204}$$

Similarly the Vm per active site for equation (200) is

$$Vm = k_c^{*}\,\frac{\sigma_{AA}^{6}}{\sigma^{4}}\,\frac{\alpha_{AB}^{3}}{\alpha_{AA}^{3}}[E]_0 \tag{205}$$

and that for equation (201) is

$$Vm' = k_c^{*}\,\frac{\sigma_{AA}}{\sigma^{2}}\,\frac{\alpha_{AB}}{\alpha_{AA}}\,\frac{\mu_B^{2}}{\mu_A^{2}}[E]_0 \tag{206}$$

The condition Vm' > Vm thus implies that

$$\frac{\mu_B^2}{\mu_A^2} > \frac{\sigma_{AA}^5}{\sigma^2} \frac{\alpha_{AB}^2}{\alpha_{AA}^2} \tag{207}$$

which is quite compatible with expression (204).

There are thus simple thermodynamic conditions that allow one to explain that within a multi-enzyme complex the kinetic behaviour of an enzyme may be completely changed. This effect may occur regardless of whether channelling between sites within the complex occurs or not.

9. General Conclusions

A number of general ideas and concepts have developed in the course of this discussion. It is now certainly of interest to try to summarize them.

Compartmentalization of enzyme reactions within the living cell represents a simple and elegant way to allow enzyme reactions, that are thermodynamically disfavoured, to occur. This implies that some enzyme reactions are coupled with transport processes across membranes and are thus vectorial processes. The synthesis of ATP in mitochondria and chloroplasts certainly represents the most obvious example of this type of process.

In stirred dilute solutions, diffusion of molecules is a fast process, and such that the enzyme velocity, that may be measured in these conditions, is under the sole control of the intrinsic properties of the enzyme. If the diffusion rate is lowered, as it must occur in living cells, the reaction rate that is determined is that of a system resulting from the coupling of diffusion and enzyme reaction. From this coupling, novel properties may emerge that are neither characteristic of diffusion, nor of enzyme reaction. The most prominent of these properties is the existence of hysteresis, that is the overall system may be able to detect not only the actual concentration of a substrate, but also whether this concentration is being reached after an increase or a decrease of this concentration. The enzyme system is then a physical model of a biosensor.

When an enzyme is embedded in a charged membrane, electrostatic repulsion or attraction of its charged substrate mimics positive or negative co-operativity. Moreover, the bound enzyme then becomes extremely sensitive to slight variations in ionic strength even if the enzyme is, by itself, insensitive to these variations. The subtlety of the enzyme response to a change of substrate concentration may be still dramatically increased by the fact that this overall response is dependent upon the spatial organization of fixed charges and enzyme molecules in the membrane.

Contrary to a common belief, the overall steady state behaviour of a metabolic network is not of necessity controlled by the kinetic properties of a single enzyme, whether it displays allosteric properties or not, but rather by a set of parameters that are specific for every enzyme involved in the pathway as well as other parameters that refer to the whole integrated system. Owing to this "molecular democracy" all the enzymes involved in a metabolic network are involved in the control of this network and any perturbation in the concentration of a reaction intermediate may reverberate through the entire pathway.

An open metabolic cycle may display sustained oscillations in homogeneous solution if at least one of the enzymes involved in this cycle involves non-linear terms. The simplest case that may be expected is the one where an enzyme is inhibited by excess substrate. Moreover if this metabolic cycle is taking place at the surface of a charged membrane, electric repulsion effects exerted by that membrane may result in the same oscillatory dynamics, even if the enzyme rate laws do not involve non-linear terms. This temporal organization of the system has been termed "dissipative structure" by Prigogine. Such dissipative structures are the consequence of a random perturbation of the cellular milieu. This is reminiscent of an old intuition of Lucretius in the *De Rerum Natura* where the author speculates that the origin of material and living objects results from a slight and random perturbation, called *clinamen*, in the fall of atoms (Prigogine and Stenger 1979).

A last degree of organizational complexity is offered when two, or more than two, enzymes catalysing different reactions are associated as an enzyme complex. Then one may show that, as a general rule, the association of different enzymes results in alterations of the behaviour of each individual enzyme.

As all these effects may be shown to occur and may be studied on simple model systems, there is little doubt that they also occur in the living cells and that spatial and functional enzyme organization is an essential aspect of enzyme functioning *in vivo*. This cannot be approached even by a complete knowledge of the kinetic, or of the dynamic, properties of individual enzymes studied in solution. Although we know much about enzyme mechanisms in solution, our knowledge of enzyme behaviour in organized systems is still scarce and requires both the use of model systems and investigations conducted in terms of cell biology.

Last but not least it may be of interest to discuss some of the results described above in the light of general ideas on the nature of scientific thought (Crombie 1959). Various philosophers and historians of science have suggested that any theory is based on preconceived ideas because these ideas are anterior to experimental results. Thus Popper (1959) for instance has outlined that it is a theory which serves as a guideline for the experiments to be done as an attempt to falsify that theory.

According to Harrisson (1987) any non-vitalistic biological theory relies upon either of the three following preconceived ideas: the belief that the macroscopic properties of living systems may be explained either through the detailed knowledge of the structure of some informational macromolecules, the attempt to apply equilibrium thermodynamics to living objects, or an assay of understanding the essence of life through kinetics and nonequilibrium thermodynamics. The first attitude is reminiscent of what Popper (1945) has called "methodological essentialism" and aims at understanding the changing properties of living systems by some "invariants", or "ideas". As outlined by Monod (1970) a gene would represent an "idea" in the platonistic sense of the term. This type of approach of living organisms is that of molecular and structural biology. The second type of approach is derived form the first, and aims at understanding the structure of biological objects, such as biomacromolecules, through equilibrium thermodynamics, by minimization of the energy of the folded polypeptide chain. The last type of preconceived idea is to apply kinetics and nonequilibrium thermodynamics so as to understand in physical terms the temporal evolution of biological properties. This is relevant of what Popper (1945) has called "methodological nominalism". This philosophical attitude which is the one followed in this review is common practice in physical sciences but has for long been rare in life sciences. It seems now to be developing in a rather exceptional way and this might correspond to a "scientific revolution" (Kuhn 1962). It is our contention that these types of approach should not be antagonistic, but should converge so as to understand the logic of life processes.

References

Aspinall GO (1980) Chemistry of cell wall polysaccharides. In *The biochemistry of Plants* Stumpf PK, Conn EE eds. 3:473-500, Academic Press, NewYork.

Barnes SJ, Weitzman PDJ (1986) Organization of citric acid cycle enzymes in a multienzyme cluster. *FEBS Lett.* 201:267-270

Beudeker RF, Kuenen JG (1981) Carboxysomes: "calvinosomes"?. *FEBS Lett* 131:269-274

Boiteux A, Hess, B, Sel'kov EE (1980) Creative functions of instability and oscillations in metabolic systems. *Curr. Top. Cell. Regul.* 17:171-203

Castellan GW (1971) Physical Chemistry. 2nd edition, Addison-Wesley, Reading, Massachussetts.

Chock PB, Rhee SG, Stadtman ER (1980a) Covalently interconvertible enzyme cascade systems. In *Methods in Enzymology*. Purich DL ed. 64:297-325, Academic Press, New York.

Chock PB, Rhee SG, Stadtman ER (1980b) Interconvertible enzyme cascades in cellular regulation. *Annu. Rev. Biochem.* 49:813-843

Cleland RE, Rayle DL (1977) Reevaluation of the effect of calcium ions on auxin-induced elongation. *Plant Physiol.* 60:709-712

Crasnier M, Noat G, Ricard J (1980) Purification and molecular properties of acid phosphatase from sycamore cell walls. *Plant Cell. Environ.* 3: 225-229

Crasnier M, Moustacas AM, Ricard J (1985) Electrostatic effects and calcium ion concentration as modulators of acid phosphatase bound to plant cell walls. *Eur J. Biochem.* 151:187-190

Crombie AC (1959) Medieval and Early Modern Science. Doubleday, New York

Darvill A, Mc Neil M, Albersheim P, Delmer D (1980) The primary cell walls of flowering plants. In *The biochemistry of Plants* Stumpf PK, Conn EE eds. 1:91-162, Academic Press, New York.

Dixon ME, Webb EC (1979) Enzymes 3rd edition, Longman London

Douzou P, Maurel P (1977a) Ionic control of biological reactions. *Trends Biochem. Sci.* 2:14-17

Douzou P, Maurel P (1977b) Ionic regulation in genetic translation systems. *Proc Natl. Acad. Sci. USA* 74:1013-1015

Dussert C, Rasigni M, Palmari J, Rasigni A, Llebaria F (1987) Minimal spanning tree analysis of biological structures. *J. theor. Biol.* 125:317-323

Dussert C, Mulliert G, Kellershohn N, Ricard J, Giordani R, Noat G, Palmari J, Rasigni M, Llebaria A, Rasigni G (1989) Molecular organization and clustering of cell-wall-bound enzymes as a source of kinetic apparent co-operativity. *Eur. J. Biochem.* 185:281-290

Engasser JM, Horvath C (1974a) Inhibition of bound enzymes. I. Antienergistic interaction of chemical and diffusional inhibition. *Biochemistry* 13:3845-3849

Engasser JM, Horvath C (1974b) Inhibition of bound enzymes. II. Characterization of product inhibition and accumulation. *Biochemistry* 13:3849-3454

Engasser JM, Horvath C (1974c) Inhibition of bound enzymes. III. Diffusion enhanced regulatory effect with substrate inhibition. *Biochemistry* 13:3855-3859

Engasser JM, Horvath C (1975) Electrostatic effects on the kinetics of bound enzymes. *Biochem J.* 145:431-435.

Engasser JM, Horvath C (1976) Diffusion and kinetics with immobilized enzymes. In *Appl. Biochem. Bioeng.* Wingard LB, Katchalski-Katzir E eds. 1:127-220, Academic Press, New York

Fell DA, Sauro MM (1985) Metabolic control and its analysis. Additional relationships between elasticities and control coefficients. *Eur. J Biochem.* 148:555-561

Fersht AR, Leatherbarrow RJ, Wells TNC (1986) Binding energy and catalysis: a lesson from protein engineering of the tyrosyl-tRNA synthetase. *Trends Biochem. Sci.* 11:321-325

Friedrich P (1985) Dynamic compartmentation in soluble multienzyme systems. In *Organized Multienzyme Systems: Catalytic Properties.* Welch GR ed. pp. 141-176, Academic Press, London

Giersch C (1988a) Control analysis of metabolic networks. 1. Homogeneous functions and the summation theorems for control coefficients. *Eur. J Biochem.* 174:509-513

Giersch C (1988b) Control analysis of metabolic networks. 2. Total differentials and general formulation of the connectivity relations. *Eur. J Biochem.* 174:515-519

Giudici-Orticoni MT, Buc J, Ricard J (1990a) Thermodynamics of information transfer between subunits in oligomeric enzymes and kinetic cooperativity. 2. Thermodynamics of kinetic cooperativity. *Eur.J. Biochem.* 194:475-481

Giudici-Orticoni MT, Buc J, Bidaud M, Ricard J (1990b) Thermodynamics of information transfer between subunits in oligomeric enzymes and kinetic cooperativity. 3. Information transfer between the subunits of chloroplast fructose bisphosphatase. *Eur. J. Biochem.* 194:483-490

Goldberg R (1984) Changes in the properties of cell wall pectin methylesterase along the *Vigna radiata* hypocotyl. *Physiol. Plant* 61:58-63

Goldberg R, Prat R (1982) Involvement of cell wall characteristics in growth processes along the mung bean hypocotyl. *Plant Cell Physiol.* 23:1145-1154

Goldbeter A, Koshland DE (1981) An amplified sensitivity arising from covalent modification in biological systems *Proc. Natl. Acad. Sci. USA* 78:6840-6844

Goldbeter A, Koshland DE (1984) Ultrasensitivity in biochemical systems controlled by covalent modification. *J. Biol. Chem.* 259:14441-14447

Goldstein L, Levin Y, Katchalsky E (1964) A water-insoluble polyanionic derivative of trypsin. II. Effect of the polyelectrolyte carrier on the kinetic behaviour of the bound trypsin. *Biochemistry* 3:1913-1919

Gontero B, Cárdenas ML, Ricard J (1988) A functional five enzyme complex of chloroplasts involved in the Calvin cycle. *Eur. J. Biochem.* 173:437-443

Hammes GG (1981) Processing of intermediate in multienzyme complex. *Biochem. Soc. Symp.* 46:73-90

Harrison LG (1987) What is the status of reaction-diffusion theory thirty-four years after Turing?. *J. theor. Biol.* 125:369-384

Heinrich R, Rapoport TA (1974) A linear steady-state treatment of enzymatic chains. Critique of the crossover theorem and a general procedure to identify interaction site with an effector. *Eur. J Biochem.* 42:97-105

Heinrich R, Rapoport SM, Rapoport TA (1977) Metabolic regulation and mathematical models. *Prog. Biophys. Mol. Biol.* 32:1-82

Hervagault JF, Thomas D (1985) Theoretical and experimental studies on the behavior of immobilized multienzyme systems. In *Organized Multienzyme Systems: Catalytic Properties.* Welch GR ed. pp. 381-418, Academic Press, New York.

Hervagault JF, Breton J, Kernevez JP, Rajani J, Thomas D (1984) Photobiochemical memory. In *Dynamics of Biochemical Systems.* Ricard J, Cornish-Bowden A eds. pp. 157-169, Plenum Press, New York.

Hess B, Kushmitz, D, Markus M (1984) Dynamic coupling and time-patterns of glycolysis. In *Dynamic of biochemical systems.* Ricard J., Cornish-Bowden A eds. pp. 213-226, Plenum Press, New York.

Hill TL (1977) Free Energy Transduction in Biology. Academic Press, New York

Hill TL, Chen Y (1970) Cooperative effects in models of steady-state transport across membranes. III. Simulation of potassium ion transport in nerve. *Proc. Natl. Acad. Sci. USA* 66:607-614

Ho CK, Fersht AT (1986) Internal thermodynamics of position 51 mutants and natural variants of tyrosyl-tRNA synthetase. *Biochemistry* 25:1891-1897

Horvath C, Engasser JM (1974) External and internal diffusion in heterogeneous enzyme systems. *Biotechnol. Bioeng.* 16:909-923

Jencks WP (1969) Catalysis in Chemistry and Enzymology. Mc Graw Hill, New York.

Jencks WP (1975) Binding energy, specificity and enzymic catalysis: The Circe effect. *Adv. Enzymol.* 43:219-410

Kacser H (1987) Control of metabolism. In *The Biochemistry of Plants.* Stumpf PK, Conn EE eds 11:39-67, Academic Press, New York

Kacser H, Burns JA (1979) Molecular democracy: Who shares the control?. *Biochem. Soc. Trans.* 7:1149-1160

Keleti T. (1984) Channelling in enzyme complexes. *In Dynamics of Biochemical Systems.* Ricard J, Cornish-Bowden A eds. pp. 103-114, Plenum Press, New York.

Keleti T, Batke K, Ovadi J, Jancsik V, Bartha F (1977) Macromolecular interactions in enzyme regulation. *Adv. Enz. Reg.* 15:233-265

Kell DB, Westerhoff HV (1985) Catalytic facilitation and membrane bioenergetics. *In Organized Multienzyme Systems: Catalytic Properties.* Welch GR ed. pp. 63-139, Academic Press, New York.

Kellershohn N, Mulliert G, Ricard J (1990) Dynamics of an open metabolic cycle at the surface of a charged membrane. I - A simple general model. *Physica D* 46:367- 379

Kirschner K, Bisswanger H (1976) Multifunctional proteins. *Annu. Rev of Biochem.* 45:143-166

Koshland DE (1962) The comparison of non-enzymic and enzymic reaction velocities. *J theor. Biol.* 2:75-86

Koshland DE, Nemethy G, Filmer D (1966) Comparison of experimental binding data and theoretical models in proteins containing subunits. *Biochemistry* 5:365-385

Koshland DE, Carraway KW, Dafforn GA, Gass JD, Storm DR (1972) The importance of orientation factors in enzymic reactions. *Cold Spring Harbor Symp. Quant. Biol.* 36:13-19

Kuhn TS (1962) The Structure of Scientific Revolutions. The University of Chicago Press, Chicago

Kurganov BI, Sugrobova NP, Mil'man LS (1985) Supramolecular organisation of glycolytic enzymes. *J. theor. Biol.* 116:509-526

Lamport DTA (1965) The protein component of primary cell walls. *Adv. Bot. Res.* 2:151-218

Lamport DTA (1980) Structure and function of plant glycoproteins. In *The Biochemistry of Plants.* Stumpf PK, Conn EE eds. 3:501-541, Academic Press, New York.

Leatherbarrow RJ, Fersht AR, Winter G (1985) Transition-state stabilization in the mechanism of tyrosyl-tRNA synthetase revealed by protein engineering. *Proc. Natl. Acad. Sci. USA* 82:7840-7844

Lienhard GE, Secemshi II, Koehler KA, Lindquist RN (1972) Enzymatic catalysis and the transition state theory of reaction rates: transition state analogs. *Cold Spring Harbor Symp. Quant. Biol.* 36:45-51

Lucretius T. (1942) De Rerum Natura. Société d'Edition "Les Belles Lettres", Guillaume Budé, Paris

Lumry R (1959) Some aspects of the thermodynamics and mechanisms of enzyme catalysis. In *The Enzymes.* Boyer PD, Lardy H, Myrbäck K eds. 1:157-231, Academic Press, New York.

Maurel P, Douzou P (1976) Catalytic implications of electrostatic potentials: The lytic activity of lysozyme as a model. *J. Mol. Biol.* 102:253-264

Mitchell P (1966) Chemiosmotic coupling in oxidative and photosynthetic phosphorylation. *Biol. Rev.* 41:445-501

Mitchell P (1968) Chemiosmotic Coupling and Energy Transduction. Glynn Research Bodmin, Cornwall.

Mitchell P (1979) Keilin's respiratory chain concept and its chemiosmotic consequences. *Science* 206:1148-1159

Mitchell P, Moyle J, Mitchell R (1979) Measurement of H^+/O quotients in mitochondria and submitochondrial vesicles. In *Methods in Enzymology.* Flischer S, Packer L eds. 55:627-640, Academic Press, New York.

Monod J (1970) Le Hasard et la Nécessité. Le Seuil, Paris

Monod J, Wyman J, Cangeux JP (1965) On the nature of allosteric transitions: A plausible model. *J. Mol. Biol.* 12:88-118

Moustacas AM, Nari J, Diamantidis G, Noat G, Crasnier M, Borel M, Ricard J (1986) Electrostatic effects and the dynamics of enzyme reactions at the surface of plant cells 2. The role of pectin methyl esterase in the modulation of electrostatic effects in soybean cell walls. *Eur. J. Biochem.* 155:191-197

Moustacas AM, Nari J, Borel M Noat G, Ricard J (1991) Pectin methylesterase, metal ions and plant cell-wall extension. The role of metal ions in plant cell-wall extension. *Biochem J.* 279:351-354

Mowbray J, Moses V (1976) The tentative identification in *Escherichia coli* of a multienzyme complex with glycolytic activity. *Eur. J. Biochem.* 66:25-36

Mulliert G, Kellershohn N, Ricard J (1990) Dynamics of an open metabolic cycle at the surface of a charged membrane. II - Multiple steady states and oscillatory behavior generated by electric repulsion effects. *Physica D* 46:380-391

Nari J, Noat G, Diamantidis G, Woudstra M, Ricard J (1986) Electrostatic effects and the dynamics of enzyme reactions at the surface of plant cells. 3. Interplay between limited cell-wall autolysis, pectin methyl esterase activity and electrostatic effects in soybean cell walls. *Eur. J. Biochem.* 155:199-202

Nari J, Noat G, Ricard J (1991) Pectin methylesterase, metal ions and plant cell-wall extension. Hydrolysis of pectin by plant cell-wall pectin methylesterase. *Biochem J.* 279:343-350

Nicholson S, Easterby JS, Powls R (1986) A single chloroplast protein with latent activities of both NADPH-dependent glyceraldehyde-3-phosphate dehydrogenase and phosphoribulokinase. *FEBS Lett* 202:19-22

Nicholson S, Easterby JS, Powls R (1987) Properties of a multimeric protein complex from chloroplasts possessing potential activities of NADPH-dependent glyceraldehyde-3-phosphate dehydrogenase and phosphoribulokinase. *Eur. J. Biochem.* 162:423-431

Nicolis G, Prigogine I (1977) Self-organization in Nonequilibrium Systems. Wiley Interscience, New York

Noat G, Crasnier M, Ricard J (1980) Ionic control of acid phosphatase activity in plant cell walls: *Plant Cell Environ.* 3:225-229

Pavlidis T (1973) Biological Oscillators: Their Mathematical Analysis. Academic Press, New York

Persson LO, Johansson G (1989) Studies of protein-protein interaction using counter-current distribution in aqueous two phase systems. Partition behaviour of six Calvin-cycle enzymes from a crude spinach (*Spinacia oleracea*) chloroplast extract. *Biochem J.* 260:1-8

Plato (1950) Oeuvres Complètes. Bibliothèque de la Pléiade, Gallimard, Paris

Popper KR (1945) The Open Society and its Enemies. Routledge, London

Popper KR (1959) The logic of Scientific Discovery. Hutchinson, London

Prigogine I, Stengers I (1979) La nouvelle Alliance. Gallimard, Paris

Reed LJ (1981) Regulation of mammalian pyruvate dehydrogenase complex by a phosphorylation-dephosphorylation cycle. *Curr. Top. Cell. Reg.* 18:95-106

Reed LJ, Pettit FH, Eley MH, Hammilton L, Collins JH, Oliver RM (1975) Reconstitution of the *Escherichia coli* pyruvate dehydrogenase complex. *Proc. Natl Acad. Sci. USA* 72:3068-3072

Reich JG, Sel'kov EE (1981) Energy metabolism of the cell. A theoretical treatise. Academic Press, London.

Ricard J (1985) Organized polymeric enzyme systems: catalytic properties. In *Organized multienzyme systems.* Welch GR ed. pp. 177-240, Academic Press, New York.

Ricard J (1987a) Dynamics of multi-enzyme reactions, cell growth and perception of ionic signals from the external milieu. *J. theor. Biol.* 123:253-278

Ricard J (1987b) Enzyme regulation. In *The Biochemistry of Plants*. Stumpf PK, Conn EE eds. 11:69-105, Academic Press, New York.

Ricard J, Cornish-Bowden A (1987) Co-operative and allosteric enzymes: 20 years on. *Eur. J. Biochem.* 166:255-272

Ricard J, Noat G (1984a) Enzyme reactions at the surface of living cells. I. Electric repulsion of charged ligands and recognition of signals from the external milieu. *J. theor. Biol.* 109:555-569

Ricard J, Noat G (1984b) Enzyme reactions at the surface of living cells. II Destabilization in the membrane and conduction of signals. *J. theor. Biol.* 109:571-580

Ricard J, Noat G (1985) Subunit coupling and kinetic co-operativity of polymeric enzymes. Amplification, attenuation and inversion effects. *J theor. Biol.* 117:633-649

Ricard J, Noat G (1986a) Electrostatic effects and the dynamics of enzyme reactions at the surface of plant cells. 1. A theory of the ionic control of a complex multi-enzyme system. *Eur. J. Biochem.* 155:183-190

Ricard J, Noat G (1986b) Subunit interaction in enzyme transition states. Antagonism between substrate binding and reaction rate. *J. theor. Biol.* 111:737-753

Ricard J, Soulié JM (1982) Self-organization and dynamics of an open futile cycle. *J. theor. Biol.* 95:105-121

Ricard J, Noat G, Crasnier M, Job D (1981) Ionic control of immobilized enzymes. Kinetics of acid phosphatase bound to plant cell walls. *Biochem J.* 195:357-367

Ricard J, Kellershohn N and Mulliert G (1989) Spatial order as a source of kinetic cooperativity in organized bound enzyme systems. *Biophys. J.* 56:477–487

Ricard J, Giudici-Orticoni MT, Buc J (1990) Thermodynamics of information transfer between subunits in oligomeric enzymes and kinetic cooperativity. 1. Thermodynamics of subunit interactions partition functions and enzyme reaction rate. *Eur. J. Biochem.* 194:463-473

Ricard J, Kellershohn N, Mulliert G. (1992) Dynamic aspects of long distance functional interactions between membrane-bound enzymes. *J. theor. Biol.* 156:1-40

Robinson JB, Srere PA (1985) Organization of Krebs tricarboxylic acid cycle enzymes in mitochondria. *J. Biol. Chem.* 260:10800-10805

Robinson JB, Inman L, Sumegi B, Srere PA (1987) Further characterization of the Krebs tricarboxylic acid cycle metabolon. *J. Biol. Chem.* 262:1786-1790

Sainis JK, Harris GC (1986) The association of D-ribulose-1,5-bisphosphate carboxylase/oxygenase with phosphoriboisomerase and phosphoribulokinase. *Biochem. Biophys. Res. Commun.* 139:947-954

Sainis JK, Merriam K, Harris GC (1989) The association of D-ribulose-1,5-bisphosphate carboxylase/oxygenase with phosphoribulokinase. *Plant Physiol* 89:368-374

Salerno C, Ovadi J, Keleti T, Fasella P (1982) Kinetics of coupled reactions catalyzed by aspartate aminotransferase and glutamate dehydrogenase. *Eur. J. Biochem.* 121:511-517

Schnakenberg J (1981) Thermodynamic Network Analysis of Biological Systems. 2nd edition, Springer, Berlin.

Secemski II, Lehrer SS, Lienhard GE (1972) A transition state analog for lysozyme. *J Biol. Chem.* 247:4740-4748

Shimakata T, Stumpf PK (1982) Fatty acid synthetase of *Spinacia oleracea* leaves. *Plant Physiol.* 69:1257-1262

Srere PA (1967) Enzyme concentrations in tissues. *Science* 158:936:937

Srere PA (1972) Is there an organization of Krebs cycle enzymes in the mitochondrial matrix? In *Energy metabolism and the regulation of metabolic processes in mitochondria*. Mehlman MA, Hanson RW eds. pp. 79-91, Academic Press, New York.

Srere PA (1985a) Organization of proteins within the mithochondrion. In *Organized Multienzyme Systems: Catalytic Properties*. Welch GR ed. pp. 1-61, Academic Press, New York.

Srere PA (1985b) The metabolon. *Trends Biochem. Sci.* 10:109-110

Srere PA (1987) Complexes of sequential metabolic enzymes. *Annu. Rev. of Biochem.* 56:89-124

Srivastava DK, Bernhard SA (1987) Enzyme-enzyme interactions in supermolecular cellular structures. *Curr. Top. Cell. Reg.* 28:1-68

Storm DR, Koshland DE (1970) A source for the special catalytic power of enzymes: Orbital steering. *Proc. Natl. Acad. Sci. USA.* 66:445-452

Storm DR, Koshland DE (1972a) An indication of the magnitude of orientation factors in esterification. *J. Am. Chem. Soc.* 94:5805-5814

Storm DR, Koshland DE (1972b) Effect of small charges in orientation on reaction rate. *J. Am. Chem. Soc.* 94:5815-5825

Taiz L. (1984) Plant cell expansion: regulation of cell wall mechanical properties. *Annu. Rev. Plant Physiol.* 61:58-63

Thomas D, Barbotin JN, David A, Hervagault JF, Romette JL (1977) Experimental evidence for a kinetic and electrochemical memory in enzyme membranes. *Proc. Natl. Acad. Sci. USA* 74: 5314-5317

Tipton KF, Dixon HBF (1979) Effects of pH on enzymes. In *Methods in Enzymology*. Purich DL ed. 63:183-234, Academic Press, New York.

Tompa P, Batke J, Ovadi J (1987) How to determine the efficiency of intermediate transfer in an interacting enzyme system? *FEBS Lett.* 214:244-248

Volpe JJ, Vagelos PR (1976) Mechanisms and regulation of biosynthesis of saturated fatty acids. *Physiol. Rev.* 56:339-417

Wakil SJ, Stoops JK, Joshi VC (1983) Hybridization studies of chicken liver fatty acid synthetase. Evidence for the participation in palmitate synthesis of cysteine and phosphopantetheine sulfhydryl groups on adjacent subunits. *J. Biol Chem.* 259:13644-13647

Walsh TP, Clarke FM, Masters CJ (1977) Modification of the kinetic parameters of aldolase on binding to the actin-containing filaments of skeletal muscle. *Biochem. J.* 165:165-167

Welch GR (1977) On the role of organized multienzyme systems in cellular metabolism: a general synthesis. *Progr. Biophys. Mol. Biol.* 32:103-191

Wells TNC, Fersht AR (1986) Use of binding energy in catalysis analyzed by mutagenesis of the tyrosyl-tRNA synthetase. *Biochemistry* 25:1881-1886

Westerhoff HV, Van Dam K (1987) Thermodynamics and Control of Biological Free-Energy Transduction. Elsevier, Amsterdam.

Whitehead E (1976) Simplifications of the derivations and forms of steady-state equations for non-equilibrim random substrate-modifier and allosteric enzyme mechanisms. *Biochem. J.* 159:449-456

Wieland F, Renner L, Verfurth C, Lynen F (1979) Studies on the multienzyme complex of yeast fatty acid synthetase. Reversible dissociation and isolation of two polypedtide chains. *Eur. J. Biochem.* 94:189-197

Wolfenden R (1969) Transition state analogues for enzyme catalysis. *Nature* 223:704-705

Microbial and Genetic Approaches to the Study of Structure-Function Relationships of Proteins

Frédéric Barras[1], Marie-Claude Kilhoffer[3], Isabelle Bortoli-German[1], Jacques Haiech[2]

1. Introduction

Two opposite attitudes have been adopted by biologists in their efforts at deciphering biological processes. The reductionist approach is based on the idea that increasing the knowledge about each component of the process studied inevitably increases the understanding of the whole process. The holistic approach aims at studying the process as a whole. The first approach is favored by biochemists and biophysicists, while the second is followed by cell biologists and systematicians. A limitation of the former is that it seldom takes into account the complexity of interactions within the broader context of the cell. An intrinsic limitation of the latter lies in its global view of events which precludes targeted modification of a cellular pathway. Fortunately, this dual view is becoming less pertinent to the description of biological investigations, and the advent of techniques of molecular biology has been particularly helpful in this regard.

"Proteins constitute the working-class molecules of the cell". Hence, understanding the way they act is a prerequisite for understanding how a cell

[1] Department of Molecular Microbiology LCB-CNRS, 31 Chemin Joseph Aiguier, 13402, Marseille, CEDEX 20, France
[2] Department of Proteins of Signal Transduction, LCB-CNRS, 31 Chemin Joseph Aiguier, 13402, Marseille, CEDEX 20, France
[3] Laboratoire de Biophysique, Faculté de Pharmacie, Université Louis Pasteur, BP 24, 67401 Illkirch CEDEX, France

functions and how life evolves. Since Anfinsen's seminal studies (Epstein 1963), it has been thought that the information needed for a protein to fold stably into a functional conformation is contained in its primary structure. This dogma has been seriously challenged by the recent discoveries of proteins that act as true folding catalysts (e.g., PDI, DsbA. . .) or as folding helpers (e.g., chaperones) (for reviews see Hubbard and Sander 1991; Georgopoulos 1992; Jaenicke 1993). However, it is currently admitted that Anfinsen's view still effectively accounts for studies using isolated *in vitro* systems, but that folding helpers often require to be invoked in studies dealing with living cells where molecular crowding may influence folding (Landry and Gierash 1991).

Basically, structure-function studies follow at least two approaches. The *numerical* approach is based on the concept that the native conformation of a protein is the most stable, i.e., the conformation with minimal potential energy. Experimental validation of this prediction is hampered by the difficulty of making an exhaustive list of all the forces that apply to a protein. Determination of an explicit form of the potential energy has been obtained by choosing a set of parameters describing the force field in which the proteins move (Van Gunsteren and Mark 1992). The choice of such a set of parameters allowed determination of the potential energy of the protein, as well as numerical computation of the dynamic trajectories of the proteins over a period of a few nanoseconds. Such movements could be measured by techniques such as time-resolved fluorescence or neutron diffraction. The *linguistic* approach is based on the concept that proteins sharing common functions should possess similarity in their primary structures. The challenge is to find different languages describing different levels of organization of the proteins (e.g., primary, tertiary structures, functional) and to define translators that enable the crossing of languages. For instance, amino acids, the elementary building units, can be grouped in similarity classes, thereby allowing the primary sequence to be rewritten in a new simpler code. This might facilitate the detection of motifs throughout functionally related proteins. These "words" can in turn be used for associating a "blind" sequence with a known function. The *linguistic* approach has become scientifically sound with the advent of both DNA sequencing technology and site-directed mutagenesis: the former expanded the set of primary sequences, hence testing the conservation of "words", while the later allowed us to test the functional significance of these "words".

In this review, our aim is to present the contribution of classical genetic and DNA recombinant techniques to some aspects of enzymology. An eukaryotic and several prokaryotic proteins will serve as examples to illustrate the use of these techniques in testing old theoretical concepts concerning protein/ligand, protein/substrate, and domain/domain interactions and protein evolution in general. Furthermore, we discuss how the use of microbial cells may allow a complementary approach towards the deciphering of structure/function relationships in protein.

2. Protein-ligand and protein-substrate interactions

2.1. The recognition problem

The bases of protein-ligand and protein-substrate interactions remain among the most attractive problems offered by enzymes. Any enzymatic activity involves the formation of an enzyme-substrate complex as an early event. Moreover, most activities are modulated by effectors, small molecules (ligands), or other proteins. In a broader context, these problems address questions of molecular recognition and specificity. Although still popular, the old lock-and-key model has for a long time been considered too simple. The protein and the ligand probably recognize each other through a sequential series of steps. An analogy would be to compare them to two ice skaters executing their movements during a competition. A successful outcome necessitates their completing the program. Similarly, a complex will be formed by only those molecules that are able to go through a series of steps. Also, this implies that complexes can be formed between "wrong" partners, their fate as a complex soon aborting at one or the other of the "movements". Therefore, to understand the rules which govern molecular recognition and specificity, we might need to gain access to the choreographic program. Time has a paramount importance in this dynamic view.

The importance of the time factor necessitates the use of techniques which allow us to explore the time scale over which recognition processes take place. It is reasonable to think that most of these "movements" occur in the picosecond up to the second range. Whereas the microsecond to second range has been accessible for 40 years, it is only recently that movements occurring in the picosecond to nanosecond range could be monitored.

Time-resolved fluorescence spectroscopy is the best-adapted technique for analyzing the picosecond to nanosecond range. However, it requires that a fluorescent probe be associated with the protein or with the ligand. For the protein, the most useful probes consist of tryptophan residues. An unambiguous signal will only be obtained if the protein under study possesses a unique tryptophan residue and, for a long time, this has represented a major limitation. However, site-directed mutagenesis allowing proteins to be engineered by either deleting or adding tryptophan constitutes an efficient way to solve the problem.

Such a strategy has been successfully employed in our study of calmodulin, a prototypical calcium-binding protein (Kilhoffer et al. 1992). Calcium serves as a second messenger in eukaryotes (Kilhoffer et al. 1983). Proteins detect changes in amplitude or frequency modulation of the cytoplasmic calcium concentration. Most of these calcium-binding proteins belong to a superfamily which has evolved from a common structure named the EF hand, composed of a 12-residue α helix, a 12-residue loop and another 12-residue α helix (Kilhoffer et al. 1983). A single isolated domain binds one calcium ion weakly (dissociation constant in the millimolar range, but the affinity reaches the micromolar range with a two-domain structure (Derancourt et al. 1978). When calcium binds, these proteins undergo a conformational modification that allows them to interact specifically with one or several target proteins (Haiech et al. 1989). As a consequence, a specific cellular response is induced. Calmodulin is able to interact with multiple target proteins and, therefore, is involved in multiple physiological pathways. Despite the large number of calmodulin-dependent proteins, the interaction between calmodulin and its targets is highly specific.

Recombinant DNA technology has been used to understand how calmodulin deciphers the calcium signal. At least four questions have to be solved which relate to the mechanism of calcium binding, the nature of the induced conformational changes, the molecular basis of the specificity, and the mechanism of complex activation. A synthetic calmodulin gene that could be modified at will has been designed and inserted in an *E. coli* expression vector to produce proteins in high quantity. Site-directed mutagenesis was then used for constructing calmodulin variants that each contain one tryptophan residue (Figure 1, Kilhoffer et al. 1988). This residue was inserted in such a way as not to perturb the conformation and the dynamic properties of the whole molecule. The fluorescence properties (steady-state and time-resolved fluorescence spectra) of five mutants were then used as signals to analyze the calcium-binding mechanism and the conformational and dynamic modification when calcium binds to calmodulin. This strategy allowed discrimination between different models

Fig. 1. C-α drawing of calmodulin showing the position of lateral chains of each tryptophan mutant

of calcium-binding mechanisms and, as pointed out previously, provided a demonstration that conformational coupling exists between the different calcium-binding domains. Moreover, a new working hypothesis based on the concept of conformation selection has emerged, as described in the next section.

2.2. Conformational selection by ligand interaction

Studies on calmodulin suggested that proteins are very flexible and travel through a set of conformations (Chabert et al. 1989). When a ligand binds, the set of conformations scanned is modified. To understand this concept, it is necessary to consider that a protein occupies regions of space-time, and that the function(s) of the protein is therefore dependent on several specific conformations and on the time during which it stays in one given conformation. Therefore, the protein should be considered not as a whole but as a set of interacting and fluctuating domains and its description should be that of the conformations and the trajectories of its atoms. Finally, what we have to describe is the region of space-time that it occupies and the modification of this region when an effector binds. The tryptophan mutant approach provides a way to analyze, in the nanosecond range, the structural changes that occur in the vicinity of the tryptophan residue, therefore allowing us to explore the local flexibility of a protein (Frauenfelder et al. 1992).

2.3. Protein dynamics and simulation

As pointed out in the introduction, it is possible to simulate the trajectories of the atoms of a protein during several nanoseconds. This simulation relies on the choice of a set of force-field parameters. On the other hand, measurement of the movement of a specific tryptophan residue can be achieved by the decay of fluorescence anisotropy (Karplus and Petsko 1990). Comparison between simulated and experimental data may lead to a refinement of the force-field parameters. These types of study have been enabled by the extensive use of tryptophan substitution mutants.

In protocols referred to above, it is clear that tryptophan should be inserted in a position that presents as few constraints as possible. Methods that allow discrimination between free and constrained positions are therefore required. Identification of such positions can be facilitated by a set of mutagenesis protocols that are now available, some of which are described below (Haiech and Sallantin 1985).

3. Protein evolution by point mutation

3.1. The suppressor approach: position-specific design of the substitution matrix

For a long time, comparison of different, but usually closely related, three-dimensional structures has been used for proposing the existence of a causal link between conservation residues, e.g., residues occupying equivalent positions in two different structures, and their importance in catalytic activity. Enzymes sharing 50% identity at the level of primary sequence have been thought to exhibit similar 3-D structure (Chothia and Lesk 1986). Search for sequence similarity has allowed definition of protein families whose members are likely to exhibit structural relatedness (Murzin and Chothia 1992). A major aim of such studies is the possibility of transferring information derived from one member to all other members of the same family. The advent of DNA technology has turned this formerly unattainable goal into a relatively facile event. Indeed, cloning and sequencing technologies have led to an exponential increase in available primary (deduced) protein sequences. Correspondingly, to identify conserved residues in a family, then to try to discover the cause of the conservation, is now a standard approach to the study of structure-function relationships of proteins. Consider a protein family encompassing plant, bacterial, and fungal enzymes: the identification of an invariant residue, say glutamate, throughout all sequences supports the idea that it might be critical for activity. The underlying reasoning is that if this glutamate were nonessential for activity, then evolution would have allowed considerable divergence, given the wide scope of organisms possessing the enzyme homologs. Hence, the conservation of a residue is interpreted to imply that this position is under strong functional constraints, therefore strong selective pressure. In contrast, conservation does not inform *per se* on the nature of these functional constraints: in the example above, speculation on the role of glutamate needs to invoke physico-chemical properties of this amino acid. In fact, subtle variations might be more informative than strict amino acid conservation. For instance, speculating on the necessity of a negative charge at this position becomes possible if a few sequences of the family contain an aspartate, instead of a glutamate residue.

An increase in the size of the sequence family can be obtained in two ways: obtaining new sequences from new organisms or systematically introducing mutations in a given sequence. DNA sequencing has facilitated the former, while site-directed mutagenesis has allowed the latter. The two approaches are complementary since the first will provide a set of sequences most likely to differ by more than one change. An example is given by cellulases, 80 sequences of which are now available, originating from bacteria, fungi, and plants. Sequence comparison has allowed them to be classified into nine families (Henrissat et al. 1989). Within each family, similarity varies from 20 to 80%. This wide range is ideal for allowing the recognition of invariant residues. Biochemical studies of cellulases has suggested acid-base catalysis as a mechanism (Sinnott 1990).

Therefore a search for invariant glutamate or aspartate residues was carried out in most families and site-directed mutagenesis was then used to test their involvement.

After defining where to direct the mutation, one needs to decide which residue to use for replacing the wild type. A classical strategy is to choose one for which the physico-chemical properties of the side chain follows closely those of the targeted residue. For instance, replacing a glutamate by a glutamine offers an ideal way for establishing the involvement of the acidic group. Playing with charge, size, or hydrophobicity criteria may be helpful for sorting out the properties required for the studies position. An ideal situation would be to mutate 19 times the position under study such that 20 variants each containing one of the 20 naturally occurring amino acids would be available for analysis. Such an approach is obviously labor-intensive and can hardly be envisioned for a position whose putative importance is suggested by the sequence comparison alone. Yet, it has recently become possible to construct easily 13 variants per position by using an old genetic trick: the suppression of nonsense codons.

In eubacteria, the amber UAG codon lacks a cognate tRNA and acts as a chain-terminating codon during protein synthesis (Gorini 1970). Therefore, if a sense codon in a gene is mutated to an amber codon, a truncated protein will be encoded by the mutant gene. This has allowed geneticists to search for secondary mutations, able to suppress the effects of the first amber mutation. An expected class contained true revertants, i.e., mutations which had mutated the amber codon back to the wild-type one. Another class contained genes encoding tRNAs. Detailed analysis revealed that, in all cases, the mutations affected the anticodon region in such a way that it could pair with a nonsense codon. An important feature was that the mutation did not prevent the tRNA from interacting and being charged by its cognate aminocyl tRNA synthase. Thus, synthesis of the polypeptide was no longer stopped at the amber codon. Instead, the residue encoded by the mutated (suppressor) tRNA was inserted at this junction. Geneticists were likewise able to select four specific suppressors, referred to as natural ones, coding for glutamine, tyrosine, serine, and leucine. This limited number was due to the number of mutations required for enabling the anticodon region to recognize an amber codon. Hence, expression of a gene carrying an amber mutation in the presence of any of the four suppressors allowed the synthesis of four proteins that each contained a different suppressing amino acid at the amber position. As early as 1979, Miller and co-workers exploited this for a preliminary structure-function analysis of the LacI repressor (Miller et al. 1979). Taking advantage of the availability of almost 90 nonsense mutations (most of which being amber) in the *lacI* gene, they were able to obtain and analyze approximately 300 altered repressors. However, *in vivo* genetic procedures limited the number of both nonsense mutations in the targeted gene and of tRNA suppressors available. The number of both increased dramatically after the advent of DNA recombinant techniques.

Site-directed mutagenesis allowed any codon of a cloned gene to be changed for an amber. Miller and coworkers (Normanly et al. 1986; Kleina et al. 1990b) synthesized new suppressor tRNA genes by annealing four to six oligonucleotides. One of the duplex contained an amber-pairing anticodon (i.e., 5'-CUA-3'), while the others more or less reconstructed the wild-type tRNA. The synthetic tRNAs were then tested for suppression efficiency and fidelity (Kleina et al. 1990b; Normanly et al. 1990). This led to the selection of eight additional different "synthetic" tRNA suppressors (inserting Ala, Cys, Glu, Phe, Gly, His, Lys, Pro) whose efficiency varies from 4 to 100% of the initiated synthesis, mainly as a function of the context of the targeted amber codon. Note that the context effect can be optimized for efficient suppression by changing the base 3' to the nonsense condon (Bossi 1985). Meanwhile, McClain and Foss (1988) were able to construct a suppressor inserting arginine. Synthetic and natural suppressor insert the expected residue in the suppressed proteins over 95% of the time, except in the case of the "glutamate suppressor" which inserts glutamate, glutamine, and threonine (or arginine) at 60, 20 and 20% respectively (Normanly et al. 1990). Therefore, use of the suppressor approach allows the set of available sequences after a single mutagenesis experiment to be multiplied by 13.

Functional tests and quantitation of the amount of suppressed protein produced should ideally be carried out simultaneously. This latter point significantly improves the protocol for at least three reasons. First, there is almost no reliable way of predicting the suppression efficiency which might vary from one suppressor to the other. The level of activity obtained in the functional tests to be weighed by the amount of protein produced for expressing the results in specific activity. Second, if the amber codon is quite close to the C-terminus, active stable truncated polypeptides might be produced. This can bias the activity attributed to the suppressor-inserted residue. This might be specially troublesome if the studied protein is oligomeric, i.e., the produced oligomer might be a mixture of suppressed full-length and truncated polypeptides whose properties might reflect more than the effect of the suppressor-inserted residue. Third, if the amber codon is close to the N-terminus, translation may be reinitiated. Last, one must be aware of naturally occurring read-through, i.e., amber codons are sometimes decoded in putatively suppressor-free contexts (see Miller 1991 for a discussion).

The suppressor approach method has been used for extending the studies on LacI repressor mentioned above (Kleina and Miller 1990a). The net result is that another 210 amber mutants were constructed by site-directed mutagenesis, amounting to 300 such mutants available and leading to the analysis of over 3000 versions of LacI (Miller 1991). The type of information which was obtained was that (1) over 60% of the positions are highly tolerant to changes and can accept any of the 13 or 14 residues without causing loss of function, while other positions were exquisitely sensitive to substitution, (2) sensitive positions lie at the N-terminus of the protein thought to be the DNA-binding domain, (3) proline was seldomly accepted even at tolerant positions.

Application of the suppressor approach to a protein whose 3-D structure was solved by crystallography revealed an important strength of the method. Indeed, Miller and co-workers were able to challenge the X-ray-deduced model of the thymidylate synthase by submitting 20 positions to the suppressor approach (Michaels et al. 1990). In particular, they found that an Arg, predicted to interact with a phosphate group, could be substituted by all 13 suppressor-inserted amino acids. Quite remarkable also was the fact that an Asp, predicted to be salt-bonded with an Arg, could be replaced by Ser, Cys, and to a reduced extent Tyr, but not by the potentially salt bond-forming Glu. Clearly, solving this apparent discrepancy should give a precise idea of the opposite constraint imposed by the longer chain of Glu as compared with the constraint of forming a salt bridge. These results clearly illustrate the necessity for backing up structural models with *in vivo* tests.

A third remarkable application of the suppressor approach was carried out by Poteete and collaborators (Rennell et al. 1991; Poteete et al. 1992), who systematically changed every codon (apart from the initiating AUG) of T4 lysozyme to an amber codon, giving rise to 2015 mutants. Testing the activity of the mutant *in vivo*, when 2% lysozyme activity was sufficient to confer a wild-type-like phenotype, revealed that 55% of the positions could tolerate any of the 13 amino acids, thereby establishing the remarkable mutability of this protein (Rennell et al. 1991). The results were then compared with 3-D models derived from crystallographic measurements (Poteete et al. 1992). First, buried positions were found highly destabilizing if occupied by charged residues. Proline proved to be the most frequently rejected residue, but this did not seem to correlate with the standard view of proline as an α-helix-destabilizing residue. Indeed the only helix which was refractory to the presence of proline was located inside the core, whereas eight other α-helices exposed to the solvent accommodated proline quite well. Hence, a provocative interpretation was that the hydrophilic character of proline, rather than its putative α-helix-destabilizing property, was detrimental to the buried helix. It is noteworthy, however, that typical α-helix-breaker behavior was observed for proline in a study on lambda repressor stability by Sauer group's (Sauer and Lim 1992). Second, salt-bridge-forming residues could be replaced by a very limited number of residues but, surprisingly, none of those exhibited potential for reconstituting the mutated salt bridge. The stabilizing effect seemed to be induced by specific features of these residues. This finding well illustrates a recurring problem in site-directed mutagenesis, namely, that amino acid substitutions are not neutral. Third, a most informative finding related to the tolerance pattern exhibited by the two acidic residues thought to participate directly in catalysis, the Glu residue displayed an exquisite sensitivity to substitutions as expected for a key catalytic residue, whereas the position occupied by Asp accommodated Cys or Ala residues, thereby calling into question the nature of its involvement in catalytic activity (see below).

As it stands, these and other studies using the suppressor approach proved to be very powerful tools for probing structure-function relationships in proteins *in vivo*. This approach does not *per se* allow a direct follow through towards biophysical methods that need large amount of highly purified proteins. The salient features of this latter are elsewhere. First, the possibility of easily substituting most amino acids at a given position allowed the relationship between loss of function and the contribution of the mutant amino acid to be immediately seen. Second, the suppressor approach allows a tolerance pattern to be drawn in a position-specific manner. This allows each position to be "read" in terms of constraint, instead of physico-chemical properties of side chains. Third, this approach demonstrates that the composition of the substitution matrix varies at each position in a protein. This had been anticipated for a long time, but the suppressor approach made its experimental investigation possible.

3.2. Interdependence of residues located in the hydrophobic core

Hydrophobic residues, when brought together away from the solvent, provide the protein with a stable, well-packed core, and this process can be one of the driving forces in the folding pathway. We discussed above how buried positions in the T4 lysozyme core were highly constrained, accommodating hydrophobic residues only. Studies in Sauer's laboratory on phage repressors revealed the existence of an interdependence between them. An original mutagenesis protocol, combining cassette mutagenesis and oligonucleotide synthesis, was developed, which potentially allowed any of the four nucleotides to be introduced at any of the codons (Reidhaar-Olson and Sauer 1988). Mutagenesis products were then tested *in vivo* and functional and/or stable variants analyzed. Assuming that any mutant could be isolated, the nonrecovery of a mutant was interpreted to be due to its instability. The core region of a protein whose 3-D structure was known, the N-terminal domain of lambda repressor, was likewise investigated (Lim and Sauer 1989). The authors defined a hydrophobic constraint which they found to be quite sufficient for forming the basic fold of the protein. Another aspect, referred to as a steric constraint, was seen in the fact that tolerance of a residue at a given position was dictated by the occurrence of particular residues at other positions. This interdependence appeared to vary as a function of the level of activity that was required for passing the selection test, thereby suggesting that the steric constraint finely tunes functionally related properties of the protein. The relationships between residues belonging to the hydrophobic core was to be more thoroughly analyzed in a separate study where these authors randomized three hydrophobic residues interacting in the lambda core to five hydrophobic amino acids (Val, Ile, Met, Phe, Leu). This allowed a study of the effects of constraints other than hydrophobicity, in particular steric properties (Lim and Sauer 1991). Seventy eight combinations of the 125 possible were obtained and approximately 70% of them were functional, demonstrating that hydrophobicity is a major determinant for constructing a functional protein. Nevertheless, properties similar to that of the wild type were exhibited by only two mutants. This uncovered the existence of stringent requirements intervening in a complex network of various steric constraints, the net outcome bearing a significant influence on the precise native structure. Interestingly, some combinations of residues were not allowed, presumably due to steric clashes. This led to the idea that in the course of the evolution of a protein, a mutation might not be retained because of a context-effect rather than because it is intrinsically incompatible with the fold of the protein. This emphasized a limit of mutagenesis studies that change one amino acid at the time. Information exists at a second level which is out of reach with such methods, and which may considerably narrow down the mutability space of a protein in the course of evolution.

3.3. Essential residues revisited

Before the advent of molecular biology techniques, one way for modifying the activity, specificity, or stability of an enzyme was chemical modification. Although this approach was littered with numerous potential artefacts, essentially due to its lack of specificity, it gave rise to the notion that some amino acids are essential for activity. These essential residues were generally thought to be part of a locally defined region of the enzyme, the catalytic (or active) site. It was anticipated that the latter was essentially a hydrophobic cleft containing residues which directly participate in the reaction and others which bind the substrate. Location of the catalytic site, identification of important residues, and detailed description of the enzymatic mechanism were then awaiting crystallographic studies of the enzyme (preferable complexed with inhibitors) and affinity labeling studies.

Application of the suppressor approach to T4 lysozyme revealed that the Asp^{20} residue, thought to be a catalytic residue, could be substituted by Cys. Biochemical characterization of the Asp → Cys mutant confirmed that it had 80% wild-type activity (Hardy and Poteete 1991). This led the authors to question the role of Asp. In catalysis, in most textbooks referring to hen-egg white lysozyme (HEW), the corresponding Asp residue is suggested to intervene directly by stabilizing a carbonium ion produced at an intermediate stage in the course of catalysis. Since Cys is unlikely to act as a nucleophile, Hardy and Poteete proposed that Asp, or Cys could transiently form an acyl-enzyme intermediate. Interestingly, such a mechanism had been ruled out after crystallographic analysis of the catalytic site of HEW, because of the distances separating the Asp residue and the substrate. While confirmation needs the Asp → Cys change to be made in the HEW lysozyme, this illustrates the contribution of site-directed mutagenesis to the deciphering of catalytic mechanism and it shows the limitations of static view of catalysis, obtained by X-ray analysis of enzymes.

Alterations of numerous positions located in the *Bacillus* barnase catalytic site revealed a link between stability and activity, most often as an inverse relationship (Meiering et al. 1992). In particular, inactivating the enzyme by changing the catalytic residues His for Ala led to a protein more stable by 0.5 kcal/mol.

Concomitant decrease in activity and gain in stability were observed with all positively charged residues intervening directly, mostly by substrate binding, in the reaction. Superficially, this contrasted with the fact that inactivating the enzyme by substituting the other catalytic residue, Glu, by Ala led to a protein destabilized by 2.3 kcal/mol. Changing another important Asp residue to Asn destabilized the enzyme by 3 kcal/mol while it was slightly more active than wild type. That the catalytic site was not optimized for stability was rationalized by the needs for the active site to undergo conformational changes and for space and available bonds to accommodate the substrate. Overall, this pointed to the existence of local strain in the active site that might be necessary for substrate binding and catalysis.

The concept of plasticity in the catalytic site has encouraged projects aimed at improving the efficiency or specificity of enzymes. Although both types of modifications were achieved by site-directed mutagenesis-based studies, one of the main lessons learned was that lack of a model for structure/function relationships in enzymes prevented modification of enzymes properties in a rationale way. This limitation was best illustrated by a text-book case, namely the project of switching trypsin specificity from positively charged to negatively charged substrates (see Branden and Tooze 1991). This was thought to be easily attained by substituting, at the bottom of the catalytic pocket, the Asp residue by a Lys residue. The mutant Asp → Lys exhibited totally unexpected behavior, since it lost specificity towards positively and negatively charged substrates altogether, whereas it acquired a wild-type-like activity towards Phe and Tyr substrates and activity greater than that of the wild type against Leu substrates.

An alternative approach was inspired by the genetist's credo which says that the only cleverness of researchers lies in the way they prepare the selecting medium. A remarkable illustration was recently provided by Beuve and Danchin (1992), who devised a genetic selection procedure allowing them to remodel the specificity of a bacterial cyclase. This was achieved by taking advantage of cAMP-dependent expression of a catabolite operon permitting *E. coli* to grow on maltose. A two-step procedure allowed the isolation of, first, an enzyme essentially devoid of adenylate cyclase but which had acquired guanylate cyclase activity, and, second, of an enzyme that had recovered wild-type level of adenylate cyclase and increased guanylate cyclase activity. Loss of discrimination resulted from changes at residues invariant among ACase and GCase. An interpretation was that conserved residues are located at substrate binding sites. If so, one must conclude that the discriminatory power of this latter is exquisitely sensitive to small structural changes.

The advantage of random mutagenesis followed by genetic selection using *E. coli* as host was highlighted by Knowles's group (Hermes et al. 1990). Starting from a sluggish triose phosphate isomerase (1000-fold reduction in activity due to a Glu to Asp mutation at the catalytic residue), they were able to select a series of second site mutations restoring a good level of activity. Here again, the selection procedure involved a good knowledge of carbon metabolism pathways in *E. coli*. Remarkably, all second-site mutations analyzed affected residues very close to or in the catalytic site. This showed that catalytic sites can adopt different structures and be equally functional. One might be tempted to speculate that enzymes contain different potential "solutions to the catalytic problem" which might be expressed in different conditions. As demonstrated by Beuve and Danchin (1992), these potential solutions may offer safeguard against mutagenesis as well as a way out for evolving towards new specificity.

3.4. What is a neutral position?

By inference from studies mentioned above, we might define a neutral position as one (1) which is not conserved throughout related sequences, (2) whose chemical modification does not affect activity or function of the protein, (3) whose tolerance pattern, as obtained by site-directed mutagenesis studies, is highly tolerant. The following example suggests that a fourth factor, i.e., context effect should be added. Calmodulin binds four calcium ions per molecule. Calcium is complexed in a 12-residue loop of which the first residue, an aspartate, and the last, a glutamate, are very important. The substitution of the latter by an alanine residue gave rise to a mutated protein able to bind only two calcium ions (Haiech et al. 1991). Surprisingly, however, this mutated protein exhibited wild-type properties when complexed to a target peptide. Hence, although it possesses all characteristics of an essential position in isolated calmodulin, the Glu position qualifies as a neutral one when studying the complexed calmodulin.

4. Protein evolution by rearrangements of combinatorial domains

4.1. Acquiring and integrating new domain(s)

Limited proteolysis of native proteins has revealed the domain organization of numerous proteins. Definitions of a structural domain range from "folding unit" to "closed globular structures well supported by a closed system of hydrogen bonds and built around a well packed interior of hydrophobic groups" (cited in Gabor Miklos and Campbell 1992). 3-D models might uncover the existence of substructural domains. Proteins often perform more than one biological activity and efforts have been made to associate a given activity with a given domain. Such a view pictures the proteins as strings of functional independent units.

Domain duplication, capture or loss, and re-use all together seem to have been instrumental in building modern proteins (Patthy 1991). Recently, Chothia estimated that most eukaryotic proteins were made of no more than 1000 domain families (Chothia 1992). This seemed quite surprising when considering the levels of specificity, diversity, and complexity exhibited by present-day proteins. One way out of this apparent paradox lies in the possibility that connection of domains did not solely result in connecting functions but created new ones, i.e., a two-domain protein would perform a third function exhibited by neither of the two individual domains and resulting from the interaction of the two of them. As a consequence, protein sequence analysis should be able to identify, on one hand, the original building blocks or domains and, on the other hand, the structural basis for the coupling between domains. While the former is part of a well-established research area, the latter remains out of reach for now. Another consequence of such a model of evolution is that a multifunctional protein might not be optimized for the individual functions it carries out but would be optimized for the ability to do them all together.

4.2. Protein secretion in Gram-negative bacteria

4.2.1. Domain recruitment as a strategy for acquiring a new function

Although still in their infancy, our studies on extracellular protein secretion in Gram-negative bacteria suggest that understanding of this process might illustrate domain/domain interactions. Our working model is the cellulase EGZ secreted by the phytopathogenic bacterium *Erwinia chrysanthemi.* A current model proposes that EGZ goes stepwise across the cytoplasmic membrane, through the periplasm, and eventually across the outer membrane (Pugsley 1993; Salmond and Reeves 1993). Crossing the inner- and outer membrane is controlled by the action of 6 *sec* and 14 *out* gene products, respectively. The *sec*-encoded pathway was intensively studied in *E. coli* and found to share many common traits with translocation across the endoplasmic reticulum membrane in eukaryotes (Pugsley 1993). One of these traits is that proteins go across membranes as non-native species, partly due to the existence of a N-terminal signal sequence, targeting the protein through the Sec pathway and eventually being cleaved off during transfer. Our study aimed at identifying and locating the targeting signal that allows EGZ to be recognized by the *E. chrysanthemi.* Out machinery and then transferred across the outer membrane.

Mature EGZ is composed of a N-terminal catalytic domain, linked via a Ser/Thr-rich linker region to a C-terminal cellulose binding domain (Py et al. 1991a). Endoglucanase activity of the former was found to be independent of the presence of the latter. Conversely, qualitative tests did not reveal an influence of the catalytic domain on the binding properties of the cellulose binding domain (CBD). Last, phenotypic tests indicated that deletion of the linker region had no effect on either enzymatic or binding activities (Py et al. 1993). This established functional independence of two domains. A naive view was that EGZ evolved by domain capture and that an additional "secretion domain" should be found. Evidence for this latter was sought by using a domain deletion approach, which had been successful in giving rise to the structural model above. Surprisingly, secretability was found to require the concomitant presence of both "cellulase" domains. Furthermore, insertion of eight residues in, or deletion of, the linker region specifically affected secretion, indicating the existence of secretion-related constraints in that region as well. Thus, the association of both domains was concluded to be a prerequisite for secretability (Py et al. 1993).

4.2.2. Domain-domain interaction as a strategy for self- and nonself-recognition

A totally unexpected finding made in recent years was that Out machineries occur in taxonomically unrelated Gram-negative bacteria (Pugsley 1993; Salmond and Reeves 1993). The level of conservation of the 13 to 14 *out* genes varies from 30 to 80% from one bacterial species to the other. This contrasted with the broad spectrum of both the bacteria considered and the proteins secreted by these pathways, e.g., cellulase in *Erwinia*, pullulanase in *Klebsiella*, exotoxin A in *Pseudomonas* or oligomeric cholera toxin in *Vibrio*. Even within a given bacterium, structurally and functionally unrelated enzymes are substrates for the same secretion pathway. For instance, the *E. chrysanthemi* Out pathway controls the secretion of (at least) five pectate lyases, two polygalacturonases, one pectin methly esterase, and one cellulase. Sequence comparison failed to reveal any common stretch that could act as a recognition/targeting signal. This is in line with the idea that the targeting signal is exposed on a folded form of the proteins. The CBD part of EGZ contains a disulfide bridge which we used as a reporter of the folding state of the protein in the course of its secretion (Bortoli-German et al. 1994). This permitted us to observe that EGZ goes across the outer membrane as a species containing a disulfide bridge, i.e., as a folded species. Similar results were obtained in studies on pullulanase secretion which, although in a different bacterium, follows a similar pathway to EGZ in *E. chrysanthemi* (Pugsley 1992). Clearly, this established remarkable differences between the folding constraints on proteins crossing inner and outer membranes.

E. carotovora is closely related to *E. chrysanthemi*; it induces the same plant disease as a consequence of the secretion of a similar battery of cell wall depolymerizing enzymes. On average, *E. carotovora* and *E. chrysanthemi* Out proteins share over 50% identity, as do *E. carotovora* and *E. chrysanthemi* secreted pectinases and cellulases. Yet we and others showed that secretion is species-specific, i.e., it does not work in heterologous system (He et al. 1991; Py et al. 1991b). This suggested that, within each bacterial cell, secreted proteins and secretion machineries coevolved very closely in order to maintain structural recognition possible despite the great diversity of secreted enzymes.

The species specificity was exploited in our quest for targeting signal in cellulases. For the sake of simplicity, let us assume that each domain contains a targeting signal and that both are under a special geometrical constraint controlled by the linker region. A simple hypothesis is that one signal confers species specificity while the other has a less differentiated function. This proved to be too naive an assumption since a hybrid cellulase, containing the *E. carotovora* cellulase catalytic domain and the *E. chrysanthemi* EGZ CBD, was secreted by neither of the two bacteria (Py and Barras, unpubl.).

This strongly suggested that the two domains cooperate for secretion, and, by inference, that they coevolved. One possibility is that they each contribute residues to the formation of a unique targeting signal. Another possibility is that there is no unique targeting signal but that EGZ and the secretion machinery recognize each other through a series of coordinated interactions involving different contact sites. Overall, this indicated that the image of EGZ as a cellulase differs from that of EGZ as a secreted protein, while it provided an example of domains fusion instrumental in giving rise to a new property.

5. Conclusions and perspectives

Proteins have evolved through a network of different constraints which might be tentatively related to (1) specific interaction with substrates and/or effectors, (2) targeting to different cell compartments, and (3) molecular density of the cell compartment they function in. One might therefore anticipate that a protein is made of residues whose roles range from specifically essential (submitted to a specific type of constraint in the course of evolution) to globally essential (submitted to a series of overlapping constraints). Consequently, the way a protein is pictured might depend upon which "property" is being examined. In this review, current evidence supporting the view of proteins as flexible entities traveling through a set of conformations has been presented. Similarly, proteins might travel through a set of functional forms whose expression depends upon time and location. Both conformational and functional landscapes might influence each other, as suggested by the fact that ligand binding modifies the set of conformations scanned by calmodulin. A combination of DNA and biophysical techniques, such as the "tryptophan mutant approach", provides a way to analyze these changes in the nanosecond range and in a defined environment. Further studies need to focus on conformational changes influenced by nonspecific factors such as molecular crowding or cellular location. In this respect, microbiology should significantly contribute to protein structure-function studies, since microbal cells provide us with a diversity of environmentally responsive "reagents" through which we can study factors influencing protein dynamics.

Acknowledgments. We thank Violaine Bonnefoy, Athel Cornish-Bowden, Vincent Méjean, Ariane Monneron, Jean-Claude Patte, George Salmond for criticism and helpful comments on the manuscript. Tony Poteete suggested the social view of proteins. M. Afshar kindly provided Figure 1. Scientific support from C3N1-CNUSC (Montpellier) is acknowledged. Work in our laboratory is supported by grants from the CNRS and the Ministère de l'Enseignement Supérieur et Recherche. I. B. is a recipient of a MESR fellowship.

References

Beuve A, Danchin A (1992) From adenylate cyclase to guanylate cyclase: mutational analysis of a change in substrate specificity. *J. Mol. Biol.* 225:933–938

Bossi L (1985) Informational suppression. In: Scaife J, Leach D, Galizzi A (eds) *Genetics of bacteria.* Academic Press, pp 49–64

Bortoli-German I, Brun E, Py B, Chippaux M, Barras F (1994) Periplasmic disulfide bond formation is essential for cellulase secretion by the plant pathogen *Erwinia chrysanthemi. Mol. Microbiol.* (in press)

Branden C, Tooze J (1991) *Introduction to protein structure.* Garland Publishing, Inc. New York, London. pp 238–241

Chabert M, Kilhoffer M-C, Watterson DM, Haiech J, Lami H (1989) Time-resolved fluorescence studies of VU-9 calmodulin, and engineered calmodulin possessing a single tryptophan residue. *Biochemistry* 28:6093–6098

Chothia C (1992) One thousand families for the molecular biologist. *Nature* 357:543–544

Chothia C, Lesk AM (1986) The relation between the divergence of sequence and structure in proteins. *EMBO J* 5:823–826

Derancourt J, Haiech J, Pechère J-F (1978) Binding of calcium by parvalbumin fragments. *BBA* 532:373–375

Epstein CJ, Goldberger RF, Anfinsen CB (1963) The genetic control of tertiary protein structure: studies with model systems. *Cold Spring Harbor Symp. Quant. Biol.* 28:439–449

Evtunshenkov AN, Shevichika VE, Fomichev YK (1987) Expression of the pectate lyase gene of *Erwinia chrysanthemi* ENA49 in cells of other representatives of the genus *Erwinia. Molekulyarna Genetika Mikrobiologiya i Virusologiya* 5:22–25

Frauenfelder H, Sligar SG, Wolynes PG (1992) The energy landscapes and motions of proteins. *Science* 254:1598–1603

Gabor Miklos GL, Campbell HD (1992) The evolution of protein domains and the organizational complexities of metazoans. *Curr. Opinion Genet. Dev.* 2:902–906

Georgopoulos C (1992) The emergence of the chaperone machines. *TIBS* 17:295–299

Gorini L (1970) Informational suppression. *Annu. Rev. Genet.* 4:107–134

Haiech J, Sallantin J (1985) Computer search of calcium binding sites in gene data bank: use of learning techniques to build an expert system. *Biochimie* 67:555–560

Haiech J, Kilhoffer M-C, Craig TA, Lukas TJ, Wilson E, Guerra-Santos L, Watterson DM (1989) Mutant analysis approaches to understanding calcium signal transduction through calmodulin and calmodulin regulated enzymes. In *Calcuim binding proteins in normal and transformed cells.* R Pochet, DEM Lawson, CW Heizmann eds. pp. 43–56, Plenum Publishing Corporation

Haiech J, Kilhoffer M-C, Lukas TJ, Craig TA, Roberts DM, Watterson DM (1991) Restoration of the calcium binding activity of mutant calmodulins toward normal by the presence of a calmodulin binding structure. *J. Biol. Chem.* 266:3427–3431

Hardy LW, Poteete AR (1991) Reexamination of the role of Asp^{20} in catalysis by bacteriophage T4 lysozyme. *Biochemistry* 30:9457–9463

He SY, Lindeberg M, Chatterjee AK, Collmer A (1991) *Erwinia chrysanthemi out* genes enable *Escherichia coli* to selectively secrete a diverse family of heterologous proteins to its milieu. *Proc. Natl. Acad. Sci. USA* 88:1079–1083

Henrissat B, Claeyssens M, Tomme P, Lemesle L, Mornon JP (1989) Cellulase families revealed by hydrophobic cluster analysis. *Gene* 81:83–95

Hermes JD, Blacklow SC, Knowles JR (1990) Searching sequence space by definably random mutagenesis: improving the catalytic potency of an enzyme. *Proc. Natl. Acad. Sci. USA* 87: 696–700

Hubbard TJP, Sander C (1991) The role of heat-shock and chaperone proteins in protein folding: possible molecular mechanisms. *Prot. Eng.* 4: 711–717

Jaenicke R (1993) Role of accessory proteins folding. *Curr. Op. Struct. Biol.* 3: 104–112

Karplus M, Petsko GA (1990) Molecular dynamics simulation in biology. *Nature* 347: 631–639

Kilhoffer M-C, Haiech J, Demaille JG (1983) Ion binding to calmodulin. *Mol. Cell. Biochem.* 51: 33–54

Kilhoffer M-C, Roberts DM, Adibi AO, Watterson DM, Haiech J (1988) Investigation of the mechanism of calcium binding to calmodulin. Use of an isofunctional mutant with a tryptophan introduced by site-directed mutagenesis. *J. Biol. Chem.* 263: 17023–17029

Kilhoffer M-C, Kubina M, Travers F, Haiech J (1992) Use of engineered proteins with internal tryptophan reporter groups and perturbation techniques to probe the mechanism of ligand-protein interactions: investigation of the mechanism of calcium binding to calmodulin. *Biochemistry* 31: 8098–8106

Kleina LG, Miller JH (1990a) Genetic studies of the *lac* repressor. XIII. Extensive amino acid replacement generated by the use of natural and synthetic nonsense suppressors. *J. Mol. Biol.* 212: 295–318

Kleina LG, Masson J-M, Normanly J, Abelson J, Miller JH (1990b) Construction of *Escherichia coli* amber suppressor tRNA genes. II. Synthesis of additional tRNA genes and improvement of suppressor efficiency. *J. Mol. Biol.* 213: 705–717

Landry SJ, Gierash LM (1991) Recognition of nascent polypeptides for targeting and folding. *TIBS* 16: 159–163

Lim WA, Sauer RT (1989) Alternative packing arrangements in the hydrophobic core of λ repressor. *Nature* 339: 31–36

Lim WA, Sauer RT (1991) The role of internal packing interactions in determining the structure and stability of a protein. *J. Mol. Biol.* 219: 359–376

McClain WH, Foss K (1988) Changing the acceptor identity of a transfer RNA by altering nucleotides in a "variable pocket". *Science* 241: 1804–1807

Meiering EM, Serrano L, Fersht AR (1992) Effect of active site residues in barnase on activity and stability. *J. Mol. Biol.* 225: 585–589

Michaels ML, Wan Kim C, Matthews DA, Miller JH (1990) *Escherichia coli* thymidylate synthase: amino acid substitutions by suppression of amber nonsense mutations. *Proc. Natl. Acad. Sci. USA* 87: 3957–3961

Miller JH (1991) Use of nonsense suppression to generate altered proteins. *Methods Enzymol* 208: 543–563

Miller JH, Coulondre C, Hofer M, Schmeissner U, Sommer H, Schmitz A, Lu P (1979) Genetic studies of the *lac* repressor. IX Generation of altered proteins by the suppression of nonsense mutations. *J. Mol. Biol.* 131: 191–222

Murzin AG, Chothia C (1992) Protein architecture: new superfamilies. *Cur. Op. Struct. Biol.* 2: 895–903

Normanly J, Masson J-M, Kleina LG, Abelson J, Miller JH (1986) Construction of two *Escherichia coli* amber suppressor genes $tRNA_{CUA}^{Phe}$ and $tRNA_{CUA}^{Cys}$. *Proc. Natl. Acad. Sci. USA* 83: 6548–6552

Normanly J, Kleina LG, Masson J-M, Abelson J, Miller JH (1990) Construction of *Escherichia coli* amber suppressor tRNA genes. III. Determination of tRNA specificity. *J. Mol. Biol.* 213: 719–726

Patthy L (1991) Modular exchange principles in proteins. *Curr. Op. Struct. Biol.* 1:351–361

Poteete AR, Rennell D, Bouvier SE (1992) Functional significance of conserved amino acid residues. *Proteins: Struct. Fonct. Genet.* 13:38–40

Pugsley AP (1992) Translocation of a folded protein across the outer membrane in *Escherichia coli. Proc. Natl. Acad. Sci. USA* 89:12058–12062

Pugsley AP (1993) The complete general secretory pathway in gram-negative bacteria. *Microb. Rev.* 57:50–108

Py B, Bortoli-German I, Haiech J, Chippaux M, Barras F (1991a) Cellulase EGZ of *Erwinia chrysanthemi*: structural organisation and importance of His98 and Glu133 residues for catalysis. *Prot. Eng.* 4:325–333

Py B, Salmond GPC, Chippaux M, Barras F (1991b) Secretion of cellulase in *Erwinia chrysanthemi* and *E. carotovora* is species-specific. *FEMS Microb. Lett.* 79:315–322

Py B, Chippaux M, Barras F (1993) Mutagenesis of cellulase EGZ for studying the general protein secretory pathway in *Erwinia chrysanthemi. Mol. Microbiol.* 7:785–793

Reidhaar-Olson JF, Sauer RT (1988) Combinatorial casette mutagenesis as a probe of the informational content of protein sequences. *Science* 241:53–57

Rennell D, Bouvier SE, Hardy LW, Poteete AR (1991) Systematic mutation of bacteriophage T4 lysozyme. *J. Mol. Biol.* 222:67–88

Salmond GPC, Reeves PJ (1993) Membrane traffic wardens and protein secretion in Gram-negative bacteria. *TIBS* 18:7–12

Sauer RT, Lim WA (1992) Mutational analysis of protein stability. *Curr. Opinion Struct. Biol.* 2:46–51

Sinnott ML (1990) Catalytic mechanisms of enzyme glycosyl transfer. *Chem. Rev.* 90:1171–1202

Van Gunsteren WF, Mark AE (1992) On the interpretation of biochemical data by molecular dynamics computer simulation. *Eur. J. Biochem.* 204:947–961

Recent Progress in Studies of Enzymatic Systems in Living Cells

Pierre M. Viallet[1]

1. Introduction: Why study the behavior of enzymes in single living cells?

Enzyme studies in test tubes have been performed for a long time, and most of our basic knowledge on enzymatic processes results from careful experiments on small amounts of ultrapurified material. Such experiments have to be conducted with great precaution to avoid any kind of injury to the sample, including denaturation or involuntary separation of the native enzyme into its different protomers.

These difficult studies have afforded us invaluable information on the biochemical and biophysical properties of enzymes. Isoenzymes which exist only in tiny amounts have been identified and their roles have been carefully specified. Modifications of the molecular environment of enzymes have demonstrated that the rate-determining step of the enzymatic reaction is not always the enzymatic transformation itself, and that either the *in situ* availability of the endogenous or exogenous substrates or the product diffusion from the reaction site may slow down the rate of the reaction.

Time-consuming systematic studies of separated enzymes belonging to a whole enzymatic system have provided data allowing reconstitution of the

[1]Laboratory of physical Chemistry, University of Perpignan, 52 Avenue de Villeneuve, 66860 Perpignan CEDEX, France

behavior of the whole system. Such a step-by-step reconstitutive process has allowed scientists to propose a complex but comprehensive biochemical model for the behavior of cells. Such information may seem sufficient as far as interest is focused only on a basic knowledge of cell biochemistry; they are obviously not sufficient for any application to medicine or enzyme engineering. Recent applications of immobilized enzymes to the production of high-value-added chemicals have shown that the practical reality may be different from the extrapolation of results obtained from experiments performed in homogeneous solutions. In the near future, genetic engineering will provide us with powerful tools for industrial and biochemical applications; but the real efficiency of these new tools must be checked before they can be widely used. Any further progress will then be conditioned by the identification of the process(es) responsible for the discrepancy between reality and expectation. It is obvious that such verification will be made easier with new equipment, designed to make the study of the behavior of enzymes in their natural environment possible. For this purpose, Nature has provided us with the best possible "test tubes": the *living cells.*

Generally speaking, any biological event in living cells is the result of a cascade of enzymatic steps in which a product of the reaction at step n acts as a substrate for the n + 1 step. Moreover, such complex systems of enzymes may also – in fact do – contain some parallel step in which the same compound may act as substrate for many parallel competing reactions. Moreover, most of these reactions involve at least two substrates – or a substrate and a cofactor. Disfunction of the whole system may result from the perturbation of any one of these steps. A further complication is that no accumulation of intermediate metabolite occurs in a well-balanced complex enzyme system.

The challenge is to identify the unbalanced step(s) responsible for the disfunction. Experiments on cell suspensions, without the means of looking at the intracellular machinery, rely only on the registration of modifications occurring in the composition of the cell culture medium. Such information consists only of data on the decrease in extracellular concentration of the substrates previously added to the culture medium and on the increasing concentration in the culture medium of the product(s) excreted by the cells, if any.

It may look more promising if information can be obtained from inside the cells. This is the case if the cells can be fed with a substrate, the intracellular concentration of which can be measured. An ideal situation occurs when every intermediate metabolite of this substrate is able to deliver a signal allowing the detection of its own intracellular accumulation. Such an accumulation will identify the unbalanced intermediate step of the whole system, though this does not necessarily mean the identification of the defective enzyme or cofactor. Such a situation can hardly be achieved, but convenient fluorescent probes may be used which allow systematic study of the different steps of the system.

We can now proceed to another level of complexity. It can be expected that all the cells belonging to the sample do not show the disfunction. It can be of interest, for prognosis or to improve the process, to be able to evaluate the ratio between the normal cells and the others. More precisely, it is necessary to be able to construct a histogram in which cells are displayed in terms of the importance of the observed imbalance. This implies the ability to record information on each living cell of the population, or at least on a statistically significant number of these cells. Thus we move from the aim of obtaining quantitative information on what occurs inside the cells to the need to obtain quantitative information on what occurs in individual cells. Obviously, such a challenge is easier to describe than to realize.

Let us consider now what happens in a single cell. Enzymes responsible for a biological event are not necessarily localized in a unique organelle. An obvious example is the pool of enzymes which are in charge of the destruction of the free radicals resulting from an oxidative shock. These enzymes are found in the cytoplasmic membrane as well as in the mitochondrial ones. Depending on the disfunction studied, it could be useful to obtain information only on the enzymes located in the mitochondrial membranes. This implies that the cell can be mapped and the information which came from the mitochondria selected. Such a goal can theoretically be achieved by using a TV camera but, at least with any commercially available equipment, at the expense of any spectral resolution and other loss of information.

This quest for more and more specific information on the events localized inside the cells has resulted in the construction of increasingly sophisticated equipment. The different modern techniques available to study the behavior of enzymes in living cells are described here in increasing order of complexity. It is a truism to say that at present, this order of complexity does not reflect the frequency of use of these methods.

In the following, examples of recent results will be given to illustrate the specific application of the different techniques of investigation and data handling used both for checking the intracellular activity of an enzymatic system and for studying the effect of modifications of the molecular microenvironment of the enzymes on their efficiency.

This presentation aims to give a review of recent progress in technology and data analysis which have provided and will provide us with information on the behavior of enzymes in their natural molecular microenvironment. To avoid time-consuming explanations, these results will be used to illustrate the ability of the new available equipment and data processing.

1.1. Some words about fluorescence

The purpose of this section is to provide the fluorescence microscopist with the basic concepts necessary to understand the advantages as well as the limits of the equipment, to determine the requirements needed to obtain the best information from his/her own apparatus, and to avoid being tempted by dazzling promises and prospects.

Like other optical properties, fluorescence appears attractive to the biologist because it is a noninvasive tool which allows the monitoring of biological events

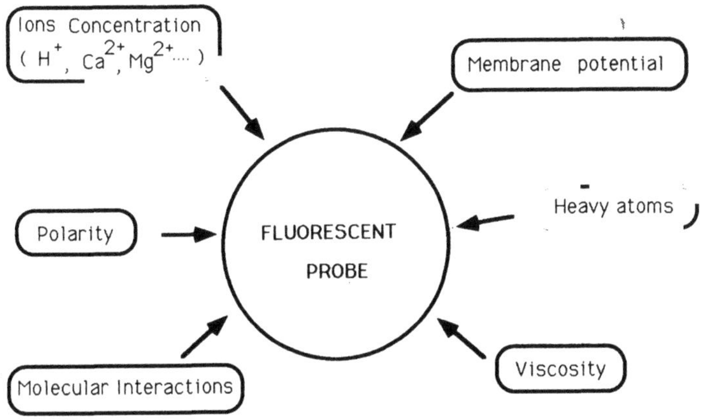

Fig. 1. Intracellular information accessible by means of fluorescent probes

in intact living cells. The main advantage of fluorescence is its high sensitivity, especially when dark-field microscopes are used.

Fluorescent compounds can be characterized by different parameters which are:

- fluorescence intensity,
- fluorescence spectrum,
- excitation spectrum,
- fluorescence lifetime,
- fluorescence polarization.

Each of these parameters is strongly dependent on the molecular environment of the fluorescent molecule, and most of them are also sensitive to its concentration, at least above some value specific for each compound. As a result, each of these parameters may provide us with valuable information not only on the local concentration of the fluorescent molecule but also on its microenvironment (Figure 1). However, reliable interpretation of the data presupposes a careful and complete study of the chemical and physical properties of the fluorescent compound.

Intensity of fluorescence

Generally speaking, the emission spectrum of a fluorescent compound seldom exhibits fine structure at room temperature. The majority of such spectra consist of a single approximatively Gaussian-shaped band, though sometimes some shoulders or secondary maxima may be observed. In the first case we need only to specify the wavelengths corresponding to the maximum and half-amplitude on both sides of the maximum to characterize the chemical. In other cases, it is better to specify also the wavelengths of shoulders and secondary maxima. From a theoretical point of view, fluorescence intensity is the sum of the energy emitted in the whole range of wavelength covered by the emission spectrum, which is visualized as the area between the baseline and the drawing of the emission spectrum. For practical purposes, fluorescence intensity is generally only evaluated at or around the maximum of the fluorescence spectrum. In both cases, the intensity is primarily dependent on the conditions of excitation which are generally fixed by the hardware of the equipment and on the concentration of the fluorescent probe. As long as the concentration remains below values specific to each fluorescent compound, there is a linear relationship between fluorescence intensity and concentration of the probe.

It is essential to point out that, contrary to what happens in solution, the local intracellular concentration of the probe is partially beyond the control of the scientist. He or she can easily monitor the extracellular concentration of the probe but its intracellular distribution depends on the physicochemical properties of the drug and organelles and on the biological status of the cells under study. For this reason, it is essential to know what kind of chemical or physical interactions may disturb the linear relationship between the concentration and the intensity of fluorescence.

It is generally known that the excess of energy accumulated by excited molecules can be eliminated by both fluorescence and phosphorescence (Figure 2). The ratio between the respective amount of energy which is released through these processes is specific to the chemical. Nevertheless, it can be modified if the microenvironment of the chemical changes during the experiment. As a first example, the presence in the vicinity of the probe of poisonous metallic ions, cadmium, lead, mercury, and platinum, will result in a decrease in its fluorescence intensity and a corresponding increase in its phosphorescence intensity through a mechanism that physicists call the "heavy atoms" effect.

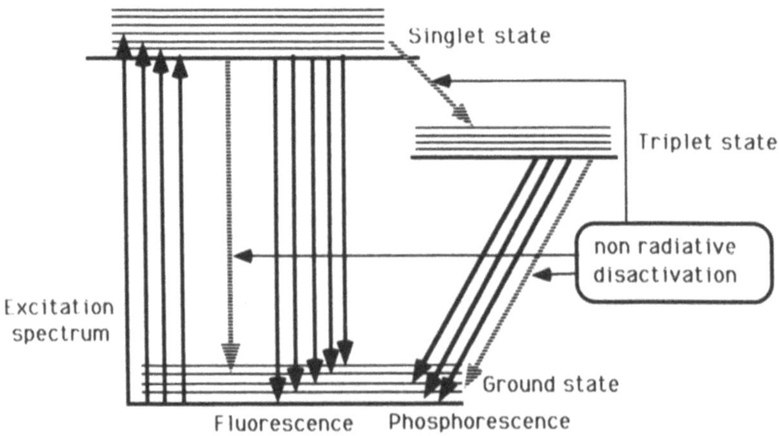

Fig. 2. Intramolecular energy transfers following electromagnetic energy absorption. *Gray arrows* represent nonradiative energy transfer

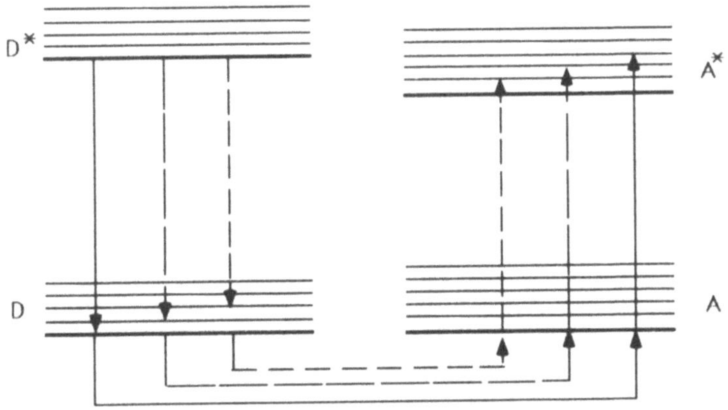

A: Long distance energy transfer

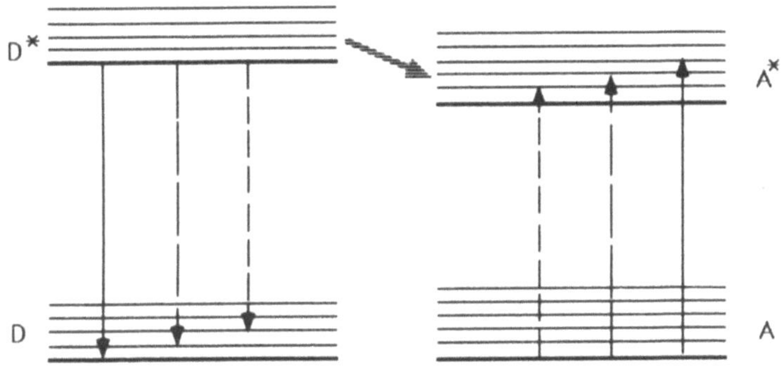

B: Resonant energy transfer

Fig. 3A, B. Intermolecular energy transfers occurring when the emission spectrum of the donor D^* partly overlaps the absorption spectrum of the acceptor A; **A** In long-distance transfer some of the radiation emitted by D^* is absorbed by the acceptor A; short distance or resonant transfer occurs when both molecules are closed enough to allow direct energy transfer between their excited states D^* and A^*. In both cases, the intensities of the same wavelengths, indicated by *dark arrows*, are affected in the same way; the *gray arrow* symbolizes the nonradiative energy transfers

More important for our purpose are the energy transfers which can occur between the excited molecule and the surrounding ones. Such energy transfers can be divided into two kinds of mechanism which can occur between both chemically identical and different molecules. Let us discuss first the case of energy transfer between different chemicals. One of these mechanisms involves the reabsorption by surrounding molecules of some amount of the light emitted by the excited molecule. Its yield is dependent on the respective position of the fluorescent and absorption spectra of the chemicals (see Figure 3A) and can occur even when the molecules are separated by rather long distances — on the cellular scale. The other mechanism of transfer, often called resonant transfer is in fact a transfer of excitation energy between two very close molecules (Figure 3B). Such a transfer may be interpreted in terms of supra-molecular interactions between the excited levels of the involved molecules. The transfer rate constant, which competes with the other rate constants of disactivation of both the donor and acceptor molecules, is dependent on the relative position of these molecules by well-established mathematical rules. It has been suggested that this kind of energy transfer, which can occur over distances as great as 8 μm, could be used as a spectroscopic ruler for the determination of distances in the 1 to 6 μM range (Föster 1948, 1949; Latt 1965; Stryer 1967; Dale 1974, 1978; Chiu 1977; Fairclough 1978; Steinberg 1978; Eisinger 1981).

Both these mechanisms always result in a decrease in the fluorescence intensity of the donor which is called molecular quenching. Both of them also affect the position and shape of the fluorescence spectrum of the acceptor. Such effects will be discussed below.

Fluorescence spectrum

As has been said above, the shape and position of a fluorescence spectrum can be used to characterize a fluorescent compound. The more structured the fluorescence spectrum, the more valuable is its use for such a characterization. Nevertheless, molecular environment may also interfere with these parameters. It has long been known that the position of an emission spectrum is dependent on the solvent polarity, and recent publications have provided us with more detailed interpretations of these shifts. It can thus be expected that careful studies of the shape and position of the fluorescence spectrum of a given fluorochrome may provide us with useful information on its molecular environment inside the cell.

Most usual fluorescent probes have small dipole moments in the ground state and larger ones in the excited state. For such a fluorochrome, the absorption spectrum is shifted to the blue when the solvent polarity increases, though the fluorescence spectrum is shifted to the red. Using the same excitation wavelength in both cases will result in both a broadening and a shift of the fluorescence spectrum to the red.

The distribution of a probe in different intracellular molecular environments with different polarities will result in a heterogeneity in the probe molecule population which, in turn, will result in a broadening of the recorded fluorescence spectrum. To make things worse, these shifts and broadenings also depend on the viscosity of the solvent, that is to say on the strength of interactions between the solvent molecules. Thus the expected shifts are representative of the heterogeneity of both the polarity and the viscosity of the probe molecular environment (Figure 4).

Nevertheless, such sensitivity to microenvironmental variations may appear interesting to use if a probe with a high degree of specificity can be designed and eventually synthetized. Let us imagine that such a probe does exist, for example

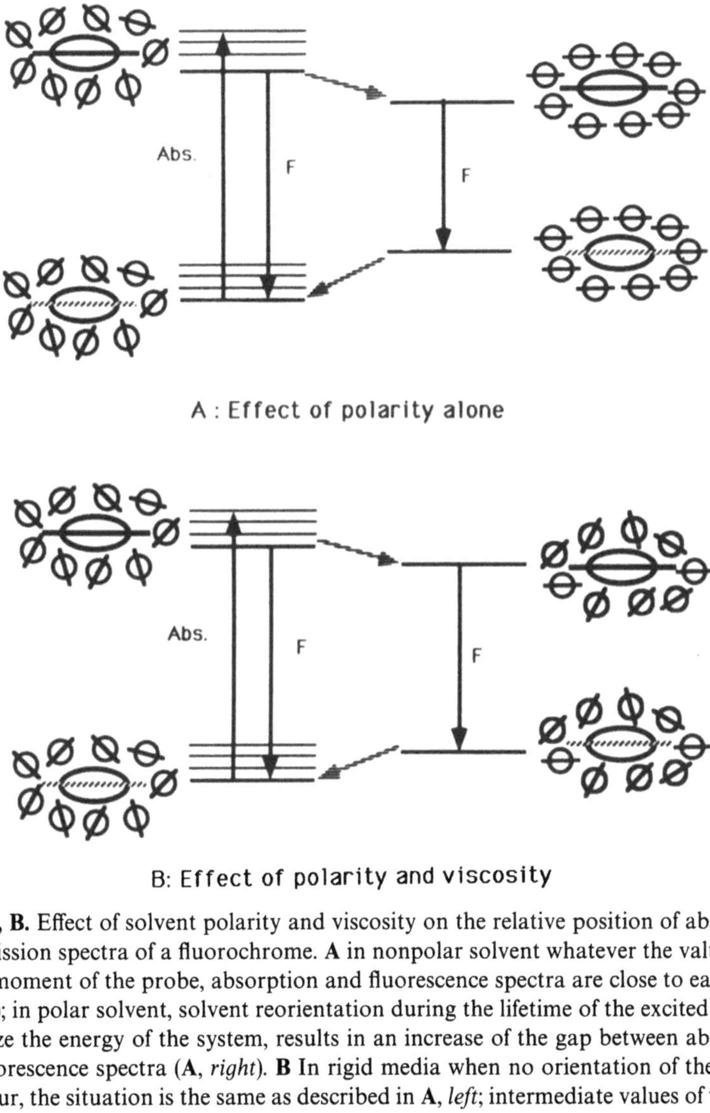

Fig. 4A, B. Effect of solvent polarity and viscosity on the relative position of absorption and emission spectra of a fluorochrome. **A** in nonpolar solvent whatever the value of the dipole moment of the probe, absorption and fluorescence spectra are close to each other (**A**, *left*); in polar solvent, solvent reorientation during the lifetime of the excited state, to minimize the energy of the system, results in an increase of the gap between absorption and fluorescence spectra (**A**, *right*). **B** In rigid media when no orientation of the solvent can occur, the situation is the same as described in **A**, *left*; intermediate values of viscosity result in situation such as described in **B**, *right Gray arrows* represent nonradiative energy transfer

for mitochondrial membranes. We cannot consider variations in energy interactions lower than one-half of the thermal energy at the experiment temperature as significant. Such a loss in energy will result in a shift to the red of about 2.5 nm of a fluorescence spectrum centered at 500 nm. These values show that we have to be very cautious in the interpretation of small spectral shifts in terms of probe membrane interactions.

More specific molecular interactions may also result in spectral shifts and/or modifications in the shape of a fluorescence spectrum. Let us come back again to energy transfer. As Figure 3 shows, both long distance and resonant energy transfer will affect only the wavelengths of the fluorescence spectrum of the donor which overlap with the absorption spectrum of the acceptor molecule. This energy transfer will induce a decrease in the donor fluorescence intensity only in the wavelength range of the overlap. Depending on the importance of this overlap, the result will be a modification of the shape of the donor fluorescence spectrum or a shift of the maximum and a modification of the shape (Weber 1989).

Autoabsorption, which occurs when a probe is too concentrated, is often relevant in long-range energy transfer. A common involuntary potential event in solution, this phenomenon is less common in experiments on living cells, due to their small dimensions. It is obvious that the probability of a photon issuing from an excited molecule being absorbed by another probe molecule is dependent on the probe concentration and on the length of the optical pathway on which the molecules are located. As long as the probe is confined inside the cell, this length cannot exceed the dimensions of the cell, and is even lower in most practical applications.

In contrast, due to the fact that fluorescent probes are generally partially lipophyllic, self-association can be expected in aqueous solutions. Their poor solubility often limits the concentration range in which fluorescent probes can be used in solution. This is a well-known situation for chemicals with a planar or quasi-planar structure, which is the case for most fluorescent probes. Such autoassociations start with the formation of dimers either in the ground state, i.e., before excitation, or in the excited state, i.e., when an excited molecule binds with an identical unexcited one. Such interactions can induce modifications of both the emission and the excitation spectra (see below). Generally speaking, this phenomenon is minimized in cells due to the dispersion of the fluorescent molecules inside the cell ultrastructures.

Excitation spectrum

With very rare exceptions, the fluorescence spectrum and quantum yield of fluorochromes dissolved in a pure solvent are independent of the excitation wavelength. Moreover, in the reduced range of wavelengths usable for microscope studies, the excitation spectrum is similar to the absorption spectrum. Such a situation is modified only when energy transfers take place between the fluorochrome and other chemicals or between different molecules of the fluorochrome.

An example of the modifications induced by autoassociation is displayed in the following graph (Morelle 1993). When the concentration of Mag-Indo, a specific probe for magnesium ions, is progressively increased in an aqueous solution, both excitation and fluorescence spectra are modified as indicated (Figure 5). It is clear from these drawings that no simple correlation between fluorescence data and the concentration of the probe can be expected when the probe concentration is as high as 10 μM.

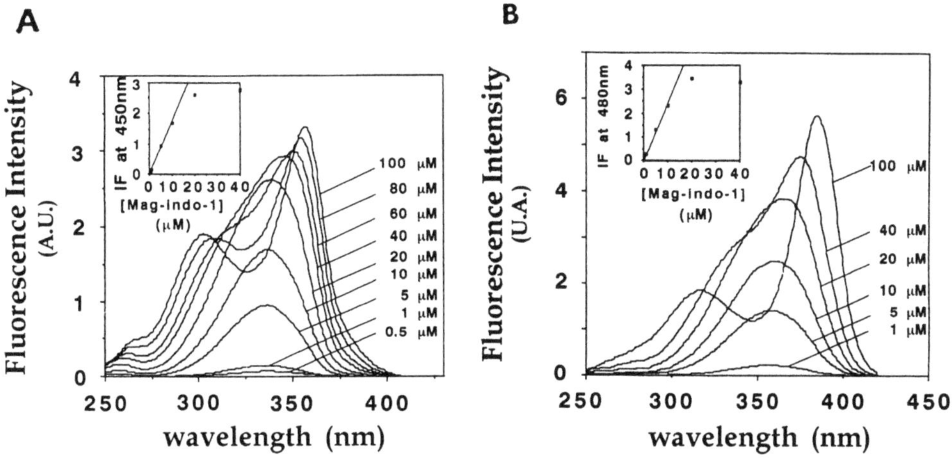

Fig. 5A, B. Influence of dimer formation on the excitation spectrum of: **A** a solution of Mag Indo in presence of a large excess of magnesium chloride; **B** a solution of Mag Indo without magnesium. Note that the recommended excitation wavelength is 345 nm. In both cases *insets* show that the linear relationship between If and the probe concentration is limited to low probe concentrations

At this point, it must again be noted that the local intracellular probe concentration is partially beyond the control of the microscopist and that some intracellular accumulation of the exogenous probe is necessary to play its role as witness of the enzyme activity. Such examples indicate the absolute necessity of having instruments sensitive enough to allow the use of very low intracellular probe concentrations.

Energy transfers between different molecules will also induce modifications in the shape of the excitation spectrum of the "acceptor". Because the "acceptor" molecule is provided with energy by the "donor" molecule, the excitation spectrum of the "acceptor" looks partially like the superposition of the "donor" absorption spectrum to its own absorption spectrum. Information coming from the excitation spectra is difficult to handle quantitatively at the cellular level, but careful examination of these spectra may be very informative in assessing a definite localization of a probe.

Finally, one has to be very cautious about potential energy transfer in the case of the simultaneous use of different intracellular fluorescent markers. Because, on the one hand, the compartimentalization of the probes is rarely total and, on the other hand, the spectral range of wavelengths able to go through the optics of a microscope is rather narrow, both short-range and long-range energy transfers have to be expected *a priori*, and tedious experiments must be carried out before assessing that they are not relevant in the experiments.

Fluorescence lifetime and quantum yield

Fluorescence quantum yield and lifetime are almost always simultaneously affected by molecular interactions, though not necessarily in a similar way.

The fluorescence quantum yield can be changed by two kinds of interactions. Those predating the excitation can be supposed to have reached equilibrium when excitation takes place. Formation of dimers in the ground state of molecules, as previously mentioned, is an example of such a situation. Depending upon the stability of these dimers in the excited state, such interaction may or may not affect the fluorescence lifetime in a significant way.

Besides such interactions, others can occur during the life of the excited state. Fortunately, most of the usual probes have a lifetime shorter than 15–20 ns, so that most of the chemical or biological events are too slow to proceed before the fluorescence process takes place. Only reactions involving fast diffusive chemicals such as protons or oxygen may really compete with this process. As a result, changes in fluorescence most often reflect interactions occurring in the ground state. Generally speaking, processes that take place at the level of the excited state change fluorescence lifetime and quantum yield in equal proportions. This is mathematically expressed by the rule

$$F^0/F = t^{0*}/t^*,$$

where F^0 and F denote respectively the fluorescence intensity in the absence and presence of the competitive process which reduces the yield, and t^{0*} and t^* the corresponding fluorescence lifetime. From the above it can be seen that considerable information on the modifications of the molecular probe environment can be obtained if relative lifetimes and intensities are measured.

Free radicals and oxygen may strongly affect both fluorescence lifetime and intensity by increasing the rate of the $S_1 \rightarrow T_1$ nonradiative energy transfer (Figure 2). Thus the production of free radicals in the vicinity of a probe can be detected through lifetime measurements. In the same way, convenient probes can be used to monitor the diffusion of oxygen molecules through cellular structures. An interesting field of application can then be found in the studies of consequences of "oxidative shock". Time resolved fluorescence spectra, i.e., fluorescence spectra recorded during the lifetime of the excited state, can sometimes be registered to provide at the same time information on the presence of free radicals or oxygen in the vicinity of the probe and on its potential influence on the probe environment (Gratton 1984; Teale 1983; Ware 1971, 1983; Weber 1984).

Fluorescence depolarization

The probability of absorption of light by a molecule depends on the angle between the direction of the electric oscillation of the light waves and a particular direction in the molecule called the absorption dipole direction. As a consequence, illumination of an ensemble of randomly distributed absorbing molecules using polarized light results in a process of photoselection, in which

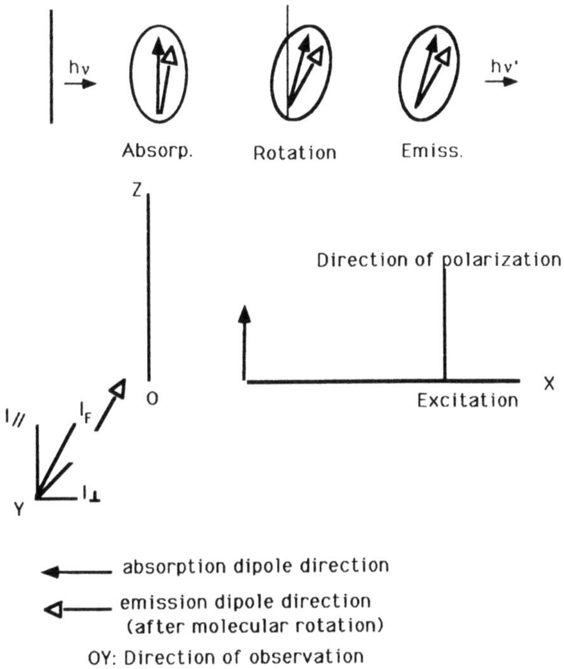

Fig. 6. Irradiation of a molecule with polarized light. If the irradiated molecule is unable to rotate, its fluorescence will be fully polarized; a partially allowed rotation will produce a partially depolarized fluorescence

the molecules with absorption dipole direction parallel to the electric field of the light are preferentially excited. Fluorescence is also associated with another particular direction in the molecule, the emission dipole direction. Usually, the emission dipole direction is parallel, or nearly so, to the direction of the absorption dipole (Figure 6). Consequently, if the fluorescent molecules do not lose their initial orientation during the fluorescence lifetime, the measured polarization of the fluorescence will reach a value specific to the chemical and related to the angle between the absorption and emission dipole. Rotational diffusion of the molecule between excitation and emission will lower this value, which can be reduced to practically zero in very fluid media (Weber 1952, 1953). Physical bases of polarization and applications to studies of complex media such as micelles, lipid bilayers, synthetic nonionic vesicles, and liquid crystals have been recently reviewed (Valeur 1993).

Fluorescence depolarization measurements, also called fluorescence anisotropy measurements, can be used to determine the rotational diffusion constant of fluorochromes embedded in different media. Relative variations in the value of this parameter can be related to differences in the respective rotational microviscosity of these different media. It is assumed that such microscopic viscosity may monitor the speed of the conformational changes of the enzyme which are supposed to occur during an enzymatic reaction, and so partially monitor the rate of the enzymatic process.

Most of the experiments involving the polarization of fluorescence have been performed to measure the microscopic viscosity of biological membranes (Stoltz 1985a, b; André 1987). In first studies, the favorite lipophylic probes were perylene and diphenylhexatriene, which are widely spread in all the lipidic cell structures so that the interpretation of the results remained critical. More recent works have been performed with probes designed to remain located in specific cell membranes and to give information on membrane fluidity at a definite depth of the bilayer (Prendergast 1981; Kuhry 1983; Muller 1987).

A very important exception is the work of Udenfriend et al. (1966) on the viscosity of the cytosol. The fluorescent polarization of N-naphthylamine previously accumulated in cells has been found to be very low, which means that the cytosolic viscosity is not appreciably different from that of water. Such a result is of prime importance in correlating studies of the intracellular behavior of cytosolic enzymes with results obtained in test tubes.

Due to a partial but important depolarization of light by the optics of microscopes (Basu 1967), polarization studies have been mainly performed on cell suspensions.

Photobleaching

Photobleaching techniques are very useful in determining the lateral diffusion coefficient of probes which mimic the behavior of the lipids constitutive to cell membranes or to probes associated with proteins embedded in these cell membranes. Unfortunately, photobleaching of probes used for fluorescence experiments may also occur involuntarily if the flux of the excitation light is not carefully controlled. If fluorescence fading is not important for scientists involved in experiments in solution, microscopists often complain about the instability of probes toward illumination. It is one of the advantages of modern equipment that exposures to light of the probes and biological materials can be reduced to the time strictly necessary to record the data. Nevertheless, the availability of more and more powerful lamps and the use of lasers as source of excitation light imply that scientists must be very cautious in controlling the flux of the wavelength used to irradiate their sample and to obey the golden rule: "the lowest, the best".

1.2. Practical consequences

Scientists planning to use any technique based upon fluorescence to study biological events on intact cells face contradictory practical constraints. First, they have to generate information, which means that a sufficient amount of probe has to be internalized in the cell(s) for a significant signal to be "seen" by the detector. On the other hand, the probe molecules must be as specific as possible to the biological structure under study, but must not modify the properties of this structure. As an example, the probes used to evaluate the rotational viscosity of any cell membrane have to be located specifically into this membrane, at a concentration large enough to deliver a measurable signal but low enough not to induce a modification of the membrane viscosity. As was said by G. Weber (1989), "the ideal probe of the dynamics must be a witness of, but not an actor in, the physiological drama".

It is rather easier in studies of enzymatic activities. It is a general experimental rule that the only way to study any kind of system in its equilibrium state is first to generate some imbalance and second to learn how the system manages to recover its balance. When the different pathways of the glycolitic enzymatic system are under study, such a return to equilibrium can be followed in living cells by monitoring the intracellular fluorescence of the reduced cofactors NADH or NADPH. In other cases, fluorescent-specific substrates of the enzymes can be used. In both cases, the fluorescent probe is both a witness to and an actor in the biological event.

Whether the imbalance is created by physical (temperature, pressure, etc.) or chemical means, it has to remain small enough to induce no dramatic damage to the enzymes. This implies that exogenous substrates have to be introduced inside the cells in only small amounts. Such a constraint matches perfectly with the general conclusions resulting from the previous discussion on the fluorescence properties.

From the above, it appears that quantitative significant data should be obtained with both low concentrations of probe and low flux of excitation. Thus, only very low levels of fluorescence intensity can be expected, which in turn implies that only very sensitive equipment can be used for such studies. We will see later that such low signals are the only ones which can be quantitatively converted into numerical data.

2. Tools available for studying the behavior of enzymes in single living cells

It is important now to indicate some differences between experiments on living cells and conventional enzymatic studies. First, the true number of molecules of the enzyme located in each cell is unknown. In the best cases, the only accessible data will give us the number of functioning molecules, i.e., those which respond to a criterium which has been tested in cuvettes. In most cases, data obtained from a single cell reflect only a mean value resulting from the respective activity of an unknown number of enzymatic molecules.

As a consequence, any crude discrepancy observed between normal cells and "pathologically suspect" ones may be interpreted in terms of either a general decrease in enzyme activity, a selective inactivation of only some molecules, or a decrease of the whole number of existing molecules. Inactivation by itself may, in turn, result from "poisoning" or inhibition of some molecules or modifications of the enzyme microenvironment (lateral or rotational viscosity, ionic concentration, etc.).

Therefore, a comprehensive understanding of the intracellular behavior of an enzymatic system supposes that information can be recorded related both to its reactivity and to some environmental properties of the enzyme.

2.1. Instrumentation

From the previous section it is clear that some fluorescence properties can be used to detect modifications of the molecular environment of the probe, whereas others are more sensitive to chemical modifications of the probe. Moreover, techniques involved to record these properties can be ranked in different ways which all have their weaknesses and strengths. In the Section 1 of this Chapter the search for more and more precise information was emphasized. For the same reason, experiments allowing global informations on cell suspensions will be first briefly discussed using fluorescence anisotropy measurements as an example.

2.1.1. Fluorescence anisotropy of cell suspension

Results of fluorescence anisotropy experiments are expressed in terms of variations of $\langle r \rangle$, a parameter defined as follow:

$$\langle r \rangle = \frac{(I_{f_{\parallel}} - I_{b_{\parallel}}) - (I_{f_{\perp}} - I_{b_{\perp}})}{I_F}$$

$$\text{with } I_F = (I_{f_{\parallel}} - I_{f_{\parallel}}) + 2(I_{f_{\perp}} - I_{b_{\perp}}),$$

where $I_{f_{\parallel}}$ is the sample signal in the direction of polarization of excitation, $I_{f_{\parallel}}$ the sample signal in the direction perpendicular to the polarization of excitation, and $I_{b_{\parallel}}$ and $I_{b_{\perp}}$ have the same meaning as "blanks". Because the fluorescence anisotropy parameter $\langle r \rangle$ is allowed to vary only in a narrow numerical range (Valeur 1993), fluorescence anisotropy has been mainly measured with suspended membranes or cells. Erythrocyte membranes have been widely studied either with equipment allowing only steady excitation (Stoltz 1985a; Bouchy 1990) or with more sophisticated apparatus in which flash lamps (Maire 1993) or pulsed lasers allowed studies of time-resolved fluorescence anisotropy.

Theoretical analysis of the data has shown that any correlation between data on fluorescence anisotropy and the rotational diffusion coefficient may be irrelevant. That is not really a surprise, for rotational diffusion coefficients have been defined at a macroscopic level and are only strictly valid in homogeneous media. In contrast, fluorescence anisotropy experiments are performed to obtain, at the molecular level, information on a heterogeneous environment. It is the use of the concept of rotational diffusion coefficients which is questionable, not the results of fluorescence anisotropy experiments. It is not the only case where macroscopic concepts have proven to be unmeaningful and confusing at the subcellular level.

Due to the lack of theoretical support, such methods can only be used for comparison of data obtained on different cell lines which are significant only if the experimental conditions were identical. With these restrictions, fluorescence anisotropy can be very informative, for it will disclose consequences of small modifications in the membrane composition. Fluorescence anisotropy has been used to study functional properties of intact leukocytes (monocytes, lymphocytes, etc.) (Inbar 1975; Berlin 1977; Johnson 1977; Donner 1989; Muller 1989), platelets (Donner 1985), and erythrocytes (Kutchai 1982) or hybridoma cells (Eyl 1992), and potential clinical applications have been investigated (Inbar 1977; Jasmin 1981; Malle 1989).

They are some contradictions in measuring fluorescence anisotropy with cell suspensions. By nature, such experiments are supposed to give information on the part of the membrane located in the vicinity of the probe. To what extent this information is representative of the characteristics of the whole membrane remains an open question even when the probe is specific to one kind of cellular membrane. Furthermore, the amount of probe internalized per cell may vary from one cell to another. Thus data collected from a cell suspension represent only a mean value which cannot reflect the cell-to-cell variability.

By contrast, cell-by-cell experiments will be much more informative. Flow cytometry techniques look very promising for such applications (Collins 1989; Donner 1990).

2.1.2. Flow cytometry

Flow cytometry is a technique so widely used that it need not be described in detail. Its main advantages are that it allows the collection of information on single cells, a dramatic progress as compared to the experiments on cell suspension, and of data from a statistically significant number of cells in a short period of time, which will not be the case for techniques which will be discussed later. Moreover, most of the flow cytometers offer the opportunity of registering different parameters simultaneously.

Conversely, each cell spends only a very short period of time in the laser beam so that the amount of "fluorescent" photons which can reach the detector is always small. As a consequence, the resulting signal is very sensitive to noises or bias. The effects of correlated noises can be minimized by summing up results obtained on a large number of cells, but such data processing may increase the effects of bias and uncorrelated noises.

Flow cytometric analysis has been applied to the study of kinetics of ligand binding and endocytosis. Ligand internalization has been followed using different methods to distinguish internalized ligands from those remaining on the cell surface: physical removal of ligands remaining on the cell surface (Murphy 1982a, b), evaluation of their amount through labeled antibodies (Sipe 1987), or specific quenching of their fluorescence by modulating the external pH (Finney 1983). After internalization, ligand receptor complexes are normally exposed to the acidic environment of prelysosomial or lysosomial structures. Ligands labeled with pH-sensitive probes can then be used to study the kinetics of lysosomial enzymes activity and the turnover of compounds normally recycled with their receptor (Muller 1980; Murphy 1984, 1986; Illinger 1990). Some interesting discussions about the relevance of the distinction between endosomes and lysosomes are also based on such experiments (Murphy 1989).

Cyanide and oxonol dyes have been used to measure membrane potentials. Physicochemical properties of these probes are well adapted to flow cytometry analysis (Chused 1988). The cyanide dyes di-O-C5 and di-O-Cu5 have excitation and fluorescence spectra suitable for conventional equipment. Their fluorescence yields increase in low-polarity environments, their high lipid/water partition coefficients increase the relative contribution of cell-bound dye to the whole fluorescence. Unfortunately, like other cyanide probes, they uncouple oxidative phosphorylation and can be toxic due to their accumulation both in the plasma membrane and in mitochondria. For these and other reasons, cyanide dyes have been found unsuitable to determine membrane potential in intact mitochondrial-containing cells. Due to the fact that they bear a negative charge, oxonol probes do not accumulate in mitochondria, and at least one of them, bis (1,3-diisopropylbarbiturate) trimethineoxonol, is said to be suitable for performing measurements on living cells and has been used in studies of alterations in membrane potential values in lymphocytes and monocytes.

Flow cytometry has also been used to monitor intracellular ion calcium in connection with measurement of membrane potentials. Unfortunately, it will be demonstrated later that the ratio method, which is the only usable one with this kind of technique, is inappropriate due to the chemical properties of the probes used.

The main concern with flow cytometric analysis is that each cell spends too short a period of time in the laser beam. This does not leave enough time to store detailed valuable information. Other techniques must be developed which would allow more detailed analysis of the cellular events.

2.1.3. Microfluorometry (spectral mode)

A schematic diagram of a microspectrofluorometer is given in Figure 7. It consists of an inverted microscope equipped with an x, y, z stepping stage that can make small (0.5 μM) repetitive steps in each direction. Each stepping motor has its own monitoring system so that they can be simultaneously activated in order to speed up the positioning of the object in the excitation beam. The (x, y, z) positions of selected objects are stored in memory of the computer so that each of them can be restored at will in the excitation beam (Allegre 1985).

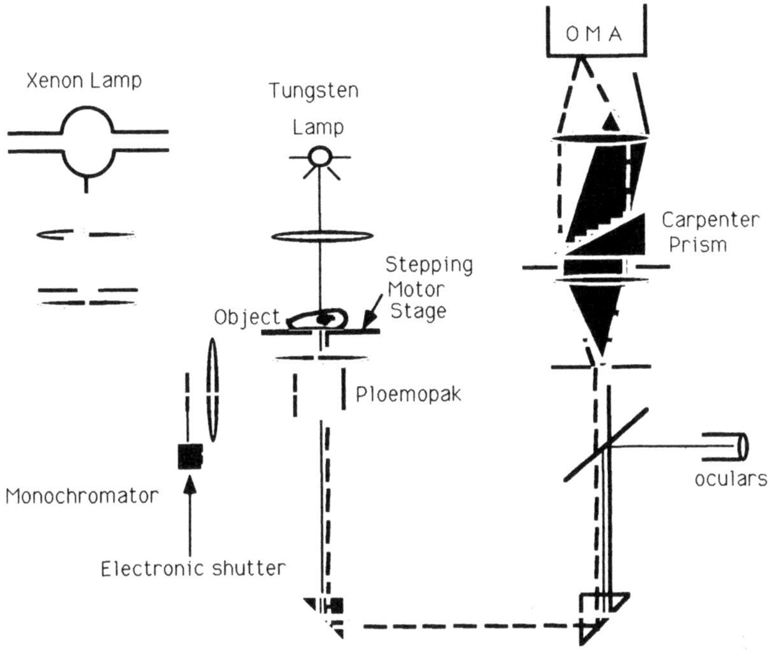

Fig. 7. Schematic diagram of a microspectrofluorometer

The monochromatic light used for the excitation is provided by a monochromator illuminated by the radiation emitted by a Xenon lamp. An electronic shutter monitored by the computer limits exposure to the time strictly necessary for data recording. An epi-illuminator can be used when needed.

The fluorescence emitted is collected by the optics of the microscope. A two-dimensional slit, located in the image plane of the microscope, allows the selection of the microscopic field to be studied. Then a miniaturized polychromator (Carpenter prism type) gives a monochromatic image of the slit on each element of the front face of the detector. The residual excitation light is emitted from this front face to avoid any potential damage.

An intensified monodimensional optical multichannel analyzer (OMA) has been selected as detector. Such a camera consists of 512 channels on which the fluorescence signal can be displayed. For practical reasons, the useful channel range is limited to about 300, which allows an optical resolution of 0.64 nm per channel.

With a SIT camera, the scanning of the 512 channels needs 64 μs, but the time lag between two scans is 32 ms. Due to the low level of fluorescence intensity, it may be necessary to sum up the signals issued from 100 scans in order to minimize the signal-to-noise ratio. This means that, for kinetic studies, a cell fluorescence spectrum can be recorded every 3 s. This will certainly result in both some damage for the cell studied and an excess of information to be stored.

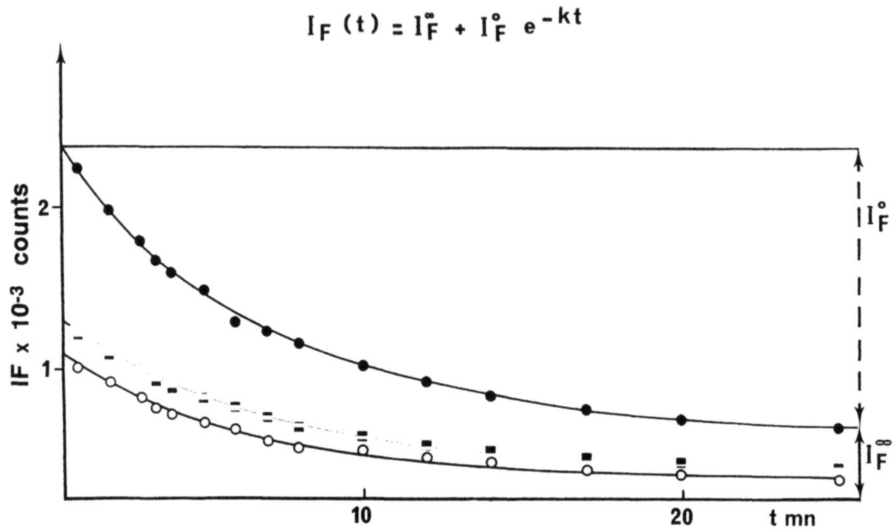

Fig. 8. An example of the decay with time of the intracellular fluorescence intensity of a 3T3 cell preloaded with B(a)P. Data shown for three different wavelengths in the range of the B(a)P fluorescence spectrum

For most of the accessible cellular enzymatic processes such as metabolism of polycyclic aromatic hydrocarbons, it is better to take advantage of this speed and of the presence of the stepping stage to study the kinetics of preselected cells at the same time. An example of the obtained decay curve is presented in Figure 8.

Stored data are then treated as explained in Section 2.2.

2.1.4. Microfluorometry (topographic mode)

Some researchers have taken advantage of the fluorescence of naturally occurring compounds such as NAD(P)H or FAD to monitor the intracellular metabolism in which they participate. NAD(P)H is usually excited with radiations in the range of 360 nm and the most convenient excitation wavelengths for the flavins are in the range of 440 nm. Because NAD(P)H and FAD cannot be excited by a good quantum yield with the same wavelength, the use of a dispersive system (Carpenter prism or grating) can be avoided. Removing the grating from the pathway of the fluorescence allows the intensity of cell fluorescence to be displayed on the channels of the OMA. Thus, each of these channels receives the fluorescence intensity issued from one particular small part of the cell, so that a readout from different sites in the cell is obtained at the same time. Data are then stored, and if such recording is repeated with time, comparative dynamic aspects of cell metabolism in different portions of the cell can be obtained.

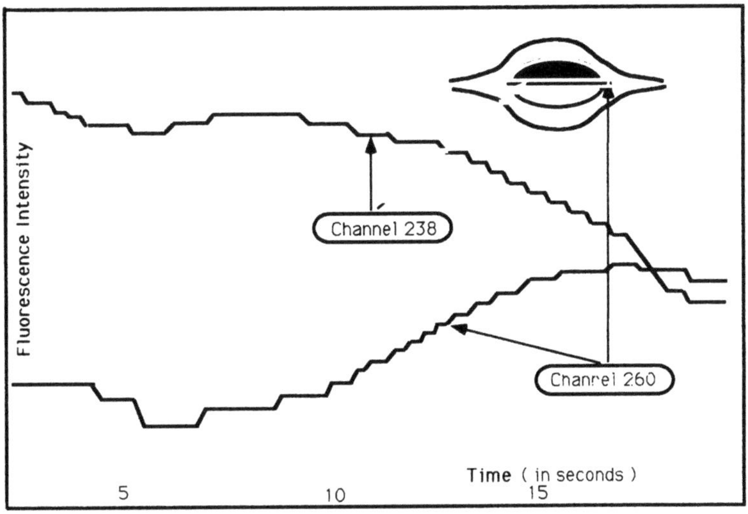

Fig. 9. Schematic representation of information accessible by means of E. Kohen equipment: increases in local NAD(P)H fluorescence after microinjection of G6P

An illustration of the above is displayed in the following scheme (Figure 9). Each curve is representative of the variation of the relative concentration of NAD(P)H in the same small volume of one cell which monitors the local ratio NAD(P)H/NAD(P). Such a ratio results from a kinetic equilibrium between the speed of biochemical events involved in the production of NAD(P)H and the speed of biochemical reactions inducing consumption of the same chemical. Thus the observed variations with time reflect the variation with time of the imbalance resulting from the microinjection of a substrate of a specific pathway of the glycolytic chain (G6P in this specific case). Although this imbalance is dependent both on the diffusion rate of the substrate from the point of microinjection to the enzymes and on the reactivity of these enzymes, such experiments provide us with the only method to monitor the *in situ* behavior of enzymes located in different parts of living cells.

The potentialities of this method are amplified if substrates specific to the different pathways can be microinjected into the cell. Comparison of the time lag in curve modifications between normal and "pathological" cells may reveal which of the pathways of the glycolytic system has been injured (Figure 10).

Dr. Elli Kohen was the initiator of these techniques and has made a life work of this difficult technology, which has explained much about metabolic control in the normal and cancerous cell (Kohen 1966; Kohen 1981a, 1986a). Microinjections of metabolic intermediates have also been used to differentiate responses of the major metabolic compartments (Kohen 1964, 1981b, 1983, 1984). The structural and functional perturbations of membrane organelles resulting

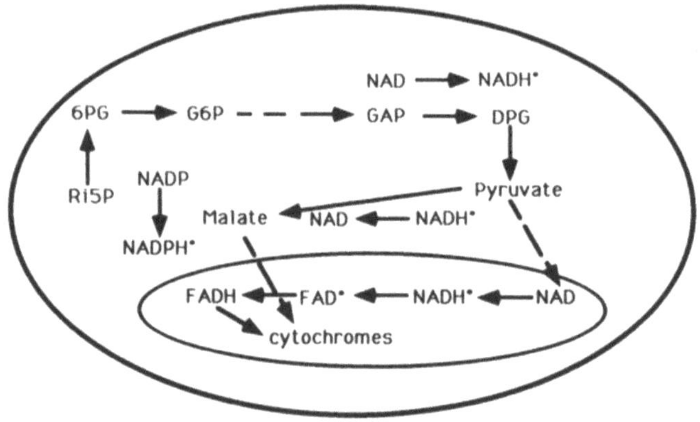

Fig. 10. Simplified scheme of the glycolytic pathway. Fluorescent compounds are labeled

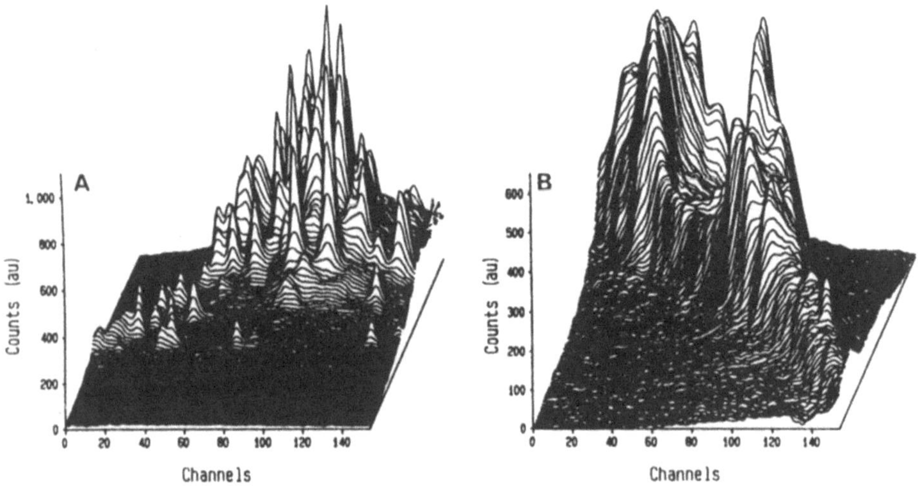

Fig. 11A, B. Topographic images of human fibroblasts incubated with: **A** TPPQ, a tetraphenylporphine linked to a chloroquinoleine ring; **B** TPP-COOH. Incubation: 2 µg/ml TPPQ or TPP-COOH (17 h) and washing (20 min) with EMEM + FCS. λexc = 435 nm, λem$>$645 nm. The field was $\cong$ 52 × 52 µm and was scanned for 5s

from drug administration have also been studied (Kohen 1986b, 1992; Blais 1992) or from photosensitization through porphyrins (Kohen 1986c; Reyftmann 1986; Moreliere 1987; Santus 1991).

As long as only unidimensional OMAs were available, the equipment had to be used in either spectral or topographic mode. Moreover, the topographic display of the probe internalization remained by nature monodimensional. Detectors with 512 · 512 channels were used by R. Santus in collaboration with

E. Kohen (Kohen 1992) to perform studies on the bidimensional mapping of the probe intracellular distribution. In this case, the orthogonal dimension are used for X-Y scans of the cell whereas the Z dimension always records the fluorescence intensity (Figure 11).

2.1.5. Videomicrofluorometry

Image analysis is being increasingly used in the biological domain, since image grabbers and image analysis systems are no longer reserved for specialists. Many departments either have their own image analyzer system or are planning to acquire one in the near future. Nevertheless, such an evolution still poses some problems for the future user, especially when he has to make a choice between an "open" system or a "closed", black-box-type system devised for a specific application. In the latter case, the experimental protocols are predefined, and the user need only execute the instructions provided by the manufacturer without being disturbed by the characteristics of the system – at least at first sight. Unfortunately, such machines can only be used with a specific probe which may be replaced in the near future by a more efficient one, a situation which cannot be considered as exceptional in such a dynamic research field. Thus the acquisition of a closed system is perfectly justified for "routine applications" when an already well-established protocol has been defined and is not susceptible to further alteration (Figure 12).

In other cases, the choice of a more open, evolutive system must be preferred; this will make it possible to follow developments in technology and will not be limiting with respect to experiment design. On the other hand, the choice of an "open" system implies that one person in the group has to dedicate himself to fully understanding the different limitations of each of the constitutive elements of the instrument (Thaer 1973; Benson 1985; Bright 1986). It is beyond the scope of this chapter to give a detailed analysis of these machines (Vigo 1991), but it is necessary to make some general remarks indicating some fundamental rules which have to be obeyed if truly quantitative information is expected.

First of all, most of these instruments have been designed to produce "beautiful" highly contrasted, sharp images, which makes them suitable for morphometric studies (Allen 1985). Unfortunately, such images are not suitable for gray-value analysis, where each pixel is supposed to have a quantitative significance (Benson 1985; Plant 1985; Choi 1988; Hause 1988; Tsunoda 1988). Let us remind the reader that a contrasted image is generally obtained by converting lightest grays into blank and darkest grays into black!

In order to give significant results, a digitized image has to be free of major defects. This is not always the case, and it is not uncommon to see digitized images containing serious faults such as image vignetting (loss of luminescence in the corners of the image), too bright backgrounds, etc., which may cause serious errors in interpretation. To achieve the necessary sensitivity, most videocameras are composed of a light intensifier coupled with a vidicon tube and an electronic interface. Photons reaching the front face of the system are

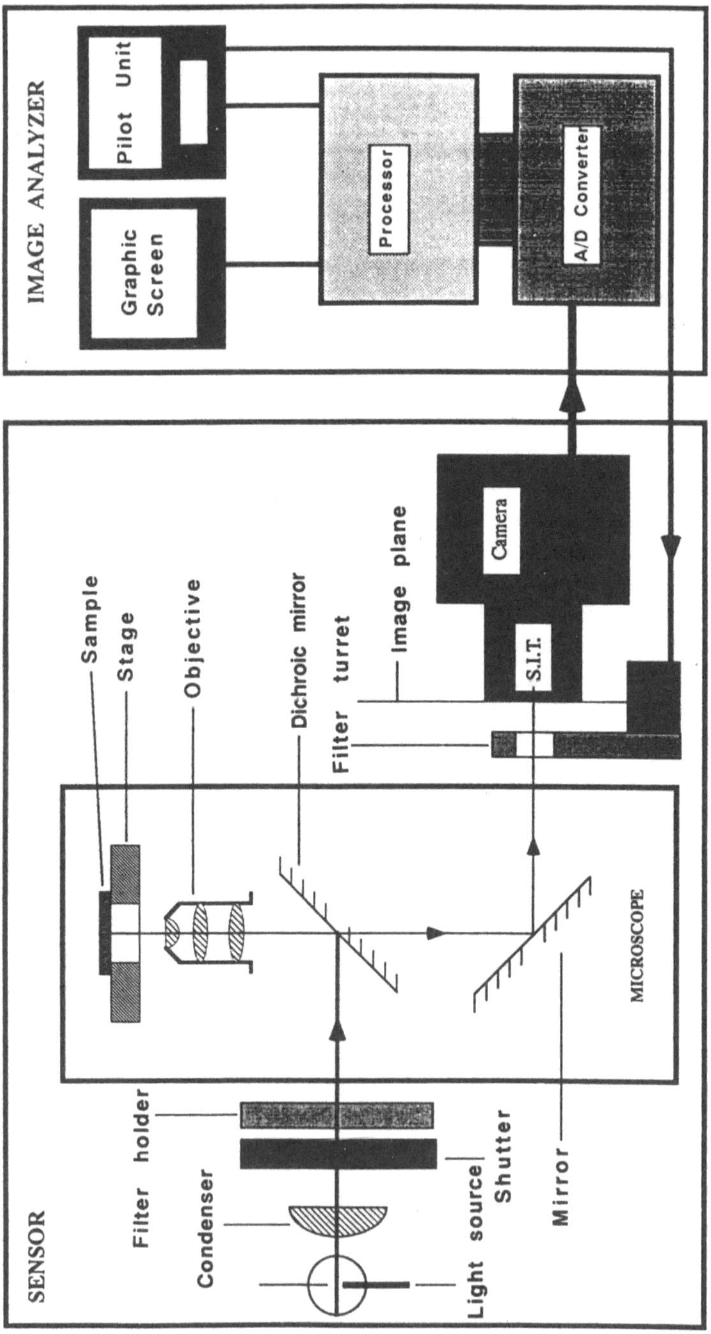

Fig. 12. Diagram of a videomicroflurometer

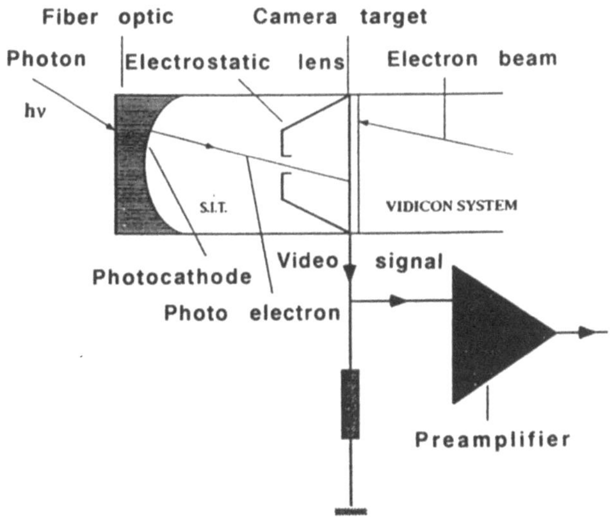

Fig. 13. SIT video camera diagram

transformed into photoelectrons. As can be seen on the diagram, due to the fact that the length of optics fibers varies according to their position, and that the angle of incidence of the photons varies with position, the yield of this transformation cannot be homogeneous (Figure 13).

This explains why some image vignetting may be observed if inadequate corrections are not performed. Such corrections are achieved by a pixel-to-pixel division of the crude image by a reference image obtained with a homogeneous lightening of the front face of the system. Note that such a division also eliminates any inhomogeneity in the yield of the mosaic of photodiodes which constitute the camera target. Of course, the reference image has to be recorded in the same conditions of use of the equipment as those used to record the crude image. An example of the importance of such a correction is illustrated by Figure 14.

Operating on images (dividing, adding, or subtracting them in order to extract the significant low signal from the unavoidable different kinds of noise in which it is buried) makes it necessary to digitize the analog electronic signal output of the camera, i.e., to quantify it into discrete numerical values. This operation is performed by an analog to digital converter interface (cf. Figure 12). Unfortunately, this conversion is not linear in the whole range of the output signal of the camera (Figure 15).

As a result, it is necessary first to minimize the noises and second to be sure to obtain low signal values on each photodiode of the camera target. Here again, the need arises to a low level of irradiation light and low local probe concentrations if quantitative information is expected. Of course, the required sensitivity

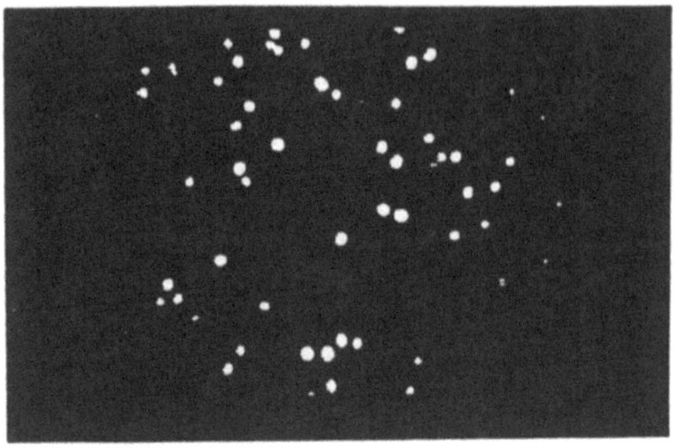

A

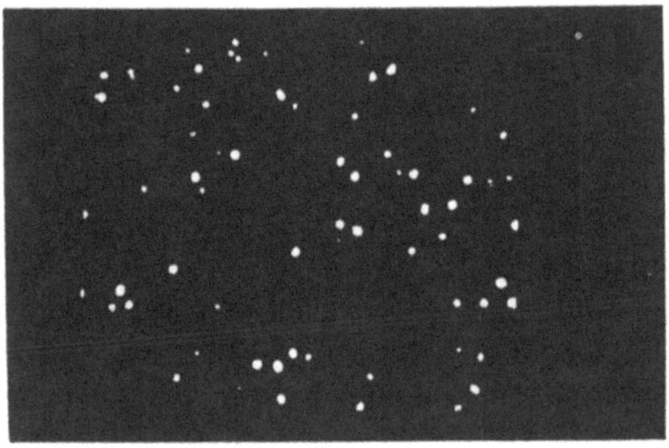

B

Fig. 14A, B. Digitized images of 3T3 fibroblast nuclei stained with HO 33342. **A** Before correction. **B** After correction

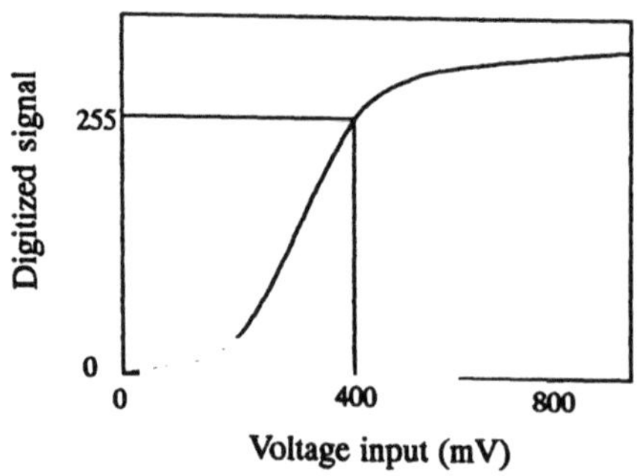

Fig. 15. Response curve of an analogical/digital convector

will be reached by the addition of successive images, a time-consuming possibility which also has its own limits.

2.1.6. Life time measurements (decay of intensity and phase fluorometry)

Subsequent to excitation into its first excited singlet state, a molecule can fluoresce or undergo various nonradiative processes which all result in the loss of the excess of absorbed energy. Each of these processes is characterized by a rate constant, as summarized below:

$$A \xrightarrow{h\nu} A^*$$

$$A^* \xrightarrow{k_1} A$$

$$A^* \xrightarrow{k_2} A,$$

so that, for an instantaneous irradiation, the decay of emission of fluorescence is as follows:

$$[A^*] = [A^*]_0 \exp\{-(k_1 + k_2)t\},$$

where $[A^*]_0$ is the concentration of $[A^*]$ when the irradiation stops. The decay constant is then $k_1 + k_2$ and the lifetime (1/e time) is:

$$\tau_{f_0} = 1/(k_1 + k_2).$$

Besides the above events, which are characteristic of the excited molecule, other mechanisms which are specific to the microenvironment of the molecule may modify τ_f, as has been previously indicated (Whal 1965; Brochon 1972; Hare 1976; Duportail 1977; Prenna 1977; Bottiroli 1979; Asins 1982; Atherton 1984; Figure 16).

$$\tau_f = 1/(k_1 + k_2 + k_3).$$

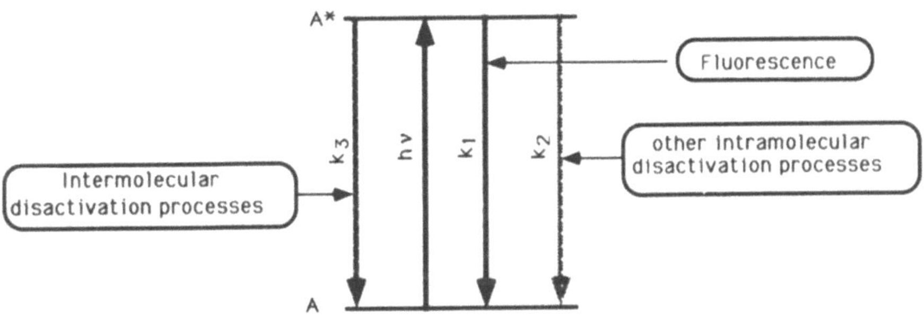

Fig. 16. Competition between intramolecular and intermolecular disactivation processes

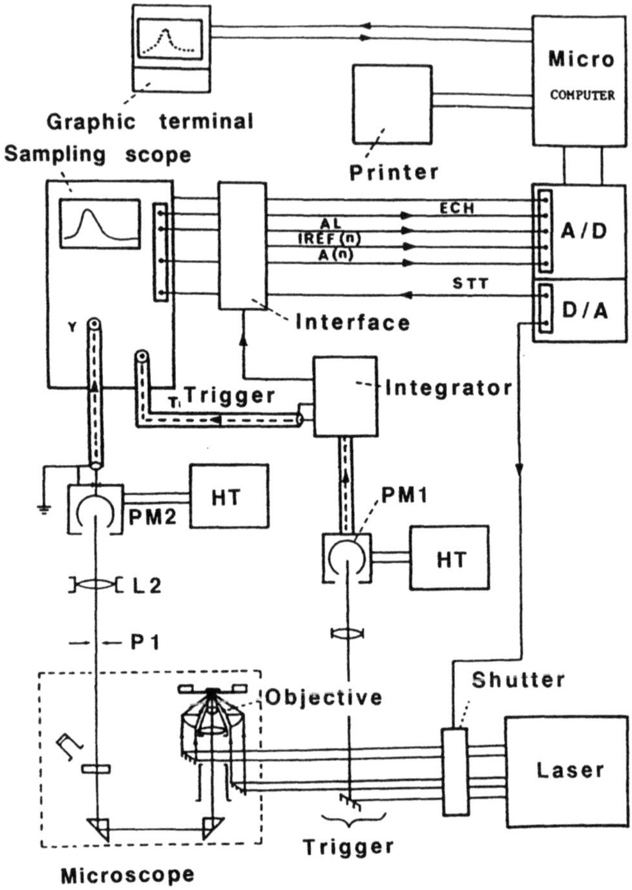

Fig. 17. Schematic representation of equipment for lifetime measurements in single living cells

Experimental determinations of τ_f can be performed through two quite distinct methods: pulse methods and phase shift techniques. Each method has its merits and advocates, but biological applications of the latter has so far been devoted to studies of the properties of isolated proteins which are beyond the scope of this chapter (Lakowicz 1989).

The availability of high-repetition-rate lasers, which provide pulse rates near 0.4 MHz, has made it possible to apply the pulse methods to studies on single living cells. With these light sources, data to recover complex decays can be obtained within minutes of data acquisition. Unfortunately, the necessary equipment for routine work is not yet commercially available.

An interesting application has nevertheless been developed using an equipment schematically described in Figure 17 (Vigo 1987). With a pulsed excitation

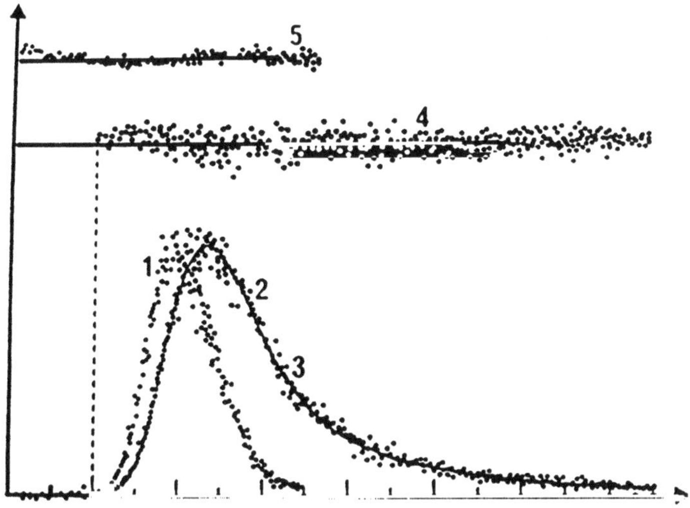

Fig. 18. Typical fluorescence response of a 3T3 cell. *1* Exciting pulse; *2* NAD(P)H observed fluorescence (*scattered points*); *3* reconvoluted theoretical curve; *4* weighted residuals; *5* autocorrelation function; λex: 360 nm

($\lambda_{exp} = 337$ nm), the fluorescence of free pyridine coenzymes in aqueous solution obeys an essentially monoexponential decay law characterized by a time constant of the order of 0.62 ns. A biexponential model generally accounts for the fluorescence of an NAD(P)H-enzyme mixture. One of the exponential corresponds to the fluorescence of free NAD(P)H, while the other, with a life time considerably longer (3–8 ns depending upon the nature of the enzyme), may be attributed to the fluorescence of enzyme-bound NAD(P)H (Scott 1970; Gafni 1976; Brochon 1976).

With this excitation wavelength, the only fluorescent cellular compounds which can be excited are the reduced forms of the nucleotides, NADH and NADPH (Kohen 1974; Salmon 1982). Applied to an isolated living cell, such a measurement should permit a quantitative evaluation of the fluorescence ratio between free and bound NAD(P)H. Experiments performed with 3T3 fibroblasts gave experimental decay curves which could not be described using a monoexponential model, whereas satisfactory fits were obtained with a biexponential decay. An example of the results obtained is shown in Figure 18. Data have been analyzed through a deconvolution program based on the modulating function method (Valeur 1973, 1978). The evaluation of the agreement between the model and the experimental signal was carried out by means of the "weighted residuals", of the variance and of the autocorrelation function (Irvin 1981; Ameloot 1982). The best fit between the experimental results (curve 2) and the reconvoluted theoretical curve (curve 3) was obtained when the following parameters were selected for the biexponential model: free NAD(P)H was found responsible for 50% of the total fluorescence signal, with a time constant lower than 1 ns, the remaining part of the recorded signal being assigned to bound NAD(P)H with a time constant of 8 ns.

Association with an enzyme also induces a shift in the fluorescence spectrum of the reduced nucleotides (Salmon 1977). Although this shift has been previously used to tentatively discriminate the free reduced nuleotides from the bound ones in living cells, pulse methods give us a new approach to this very important problem. More work has to be done in this direction to improve the sensitivity of these methods in order to allow studies on the variations of the free/bound NAD(P)H ratio after some microinjection of substrates for the glycolytic chain.

Another interesting result was obtained with cells incubated with benzo(a)pyrene. In deoxygenated solvents like nonane, B(a)P is said to have a lifetime in the range of 40–45 ns, a value which falls to about 11 ns for solvents containing atmospheric oxygen. Pulse experiments have revealed that a biexponential model accounted for the fluorescence decay of B(a)P loaded in living 3T3 cells. Furthermore, the portion of signal belonging to the short lifetime ($\cong$ 10 ns) decreased to zero during the metabolism of B(a)P so that, at the end of the experiment, the decay appeared monoexponential with a time constant of about 35 ns. Such results, which suggest that two kinds of microenvironment exist for B(a)P, are consistent with the following observations: during microspectrofluorometric experiments the contribution of the B(a)P fluorescence spectrum to the whole cell fluorescence spectrum never falls to zero, which suggests that one part of the intracellular loaded B(a)P cannot undergo metabolism (cf. Figure 8).

2.2 Progress in software and modelization

2.2.1. Spectral resolution

Rationale. Except in the case of studies on variations in NAD(P)H concentration in the absence of exogenous fluorescent chemical, a cell fluorescence spectrum results generally from the contribution of more than one fluorescent compound. Due to the broadness of the excitation spectrum of pyridine nucleotides, a cell fluorescence spectrum cannot *a priori* be assumed to be due only to the contribution of the added fluorescent probe. This is of particular importance when low intracellular probe concentrations are used, as is necessary for quantitative applications. An example of this assessment is given in Figure 19, in which the fluorescence spectrum of a benzo(a)pyrene solution is compared to the cell fluorescence spectrum of a 3T3 cell loaded with benzo(a)pyrene. Of course, when the probe is ionic, such as a pH-sensitive one, any cell fluorescence spectrum may have three components: the NAD(P)H fluorescence, often named "intrinsic cell fluorescence", the free ionized probe fluorescence, and the fluorescence of the probe associated with the ion, the protonated probe in the case of

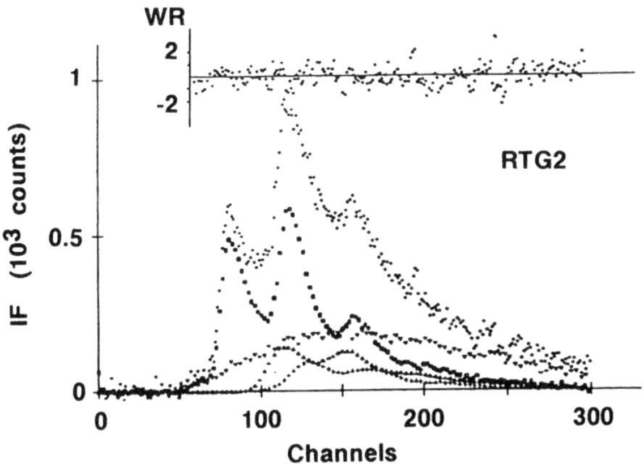

Fig. 19. Comparison between the fluorescence spectrum of B(a)P in solution (•) and of the fluorescence spectrum of a 3T3 cell loaded with B(a)P (□); (+) intrinsic cell fluorescence

pH determination (Alibaud 1988; Lahmy 1989). Such a situation precludes careless use of the popular "ratio method", which consists in dividing the fluorescence intensity recorded at a wavelength "specific" to one form of the probe by the fluorescence intensity recorded at a wavelength "specific" to the other one. In the opinion of many scientists, such a situation limits the use of fluorescent probes. The "ratio method" is so easy to use and has been claimed to have so many advantages that its rejection is associated with a rejection of the use of fluorescence techniques. Sometimes the requirement of on-line results is also put forward as a reason to neglect the contribution of the cell intrinsic fluorescence.

It is our opinion that significant information is more important than on-line, and that the amount of participation of the cell autofluorescence in the whole cellular spectrum may sometimes be used as a criterium of the amount of internalized probe. Other things being equal, an insignificant participation of the intrinsic cell fluorescence spectrum to the whole cell fluorescence may be considered as a signal of either an excess of probe inside the cell or undesired interference by the probe in the cell machinery.

The need for a convenient method of resolution of cell fluorescence spectra is still more obvious when a probe is used as a substrate for the enzyme system studied. As has been said before, such an approach takes advantage of the fact that the fluorescent probe may undergo chemical changes at each step of the whole enzymatic process. If fluorescence spectral changes are associated with these chemical modifications, identification and quantification of the different fluorescent metabolites will lead the way to simultaneous dynamic studies of the different steps involved in the whole process. However, such identifications through characteristic fluorescence spectra suppose that the intracellular ones are not spoiled by the NAD(P)H fluorescence. Of course, such a method has to be specifically designed to be used with a low level of fluorescence intensities.

Numerical resolution of fluorescence spectra. (Salmon 1988). If a complex fluorescence spectrum results from the fluorescence of N chemicals, the intensity at each wavelength can be described as a linear combination of the intensities of each component at the same wavelength. This can be expressed as follows:

$$IT_\lambda = \sum_{i=1}^{N} a_i I_{i,\lambda}, \tag{1}$$

where IT_λ is the intensity of experimental spectrum at the wavelength λ, $I_{i,\lambda}$ is the intensity of the ith characteristic spectrum, and a_i is a coefficient proportional to its contribution in the mixture. When a fluorescence spectrum emitted from a single cell is recorded, the amount of photons collected per frame scan ($32\ 10^{-3}$ s) for each channel is low. The signal-to-noise ratio can be increased by (1) increasing the number of accumulations of the signal; (2) integration of the fluorescence spectrum. The first process is time-consuming (3 s for the summation of 100 fluorescence spectra), while the second leads to the loss of spectral information. An integration of equation (1) over the whole spectral domain gives:

$$\sum_{\lambda=\lambda_0}^{\lambda_1} IT_\lambda = \sum_{i=1}^{N} a_i \sum_{\lambda=\lambda_0}^{\lambda_1} I_{i,\lambda}. \tag{2}$$

A set of N independent equations is then generated from equation (2) using N modulating functions, depending on λ, in order to restore the spectral information. Theoretically, any nonrandom function $\Phi_{j,\lambda}$ is suitable for this purpose. The originality of the method we have developed lies in (1) the use of the characteristic spectrum $I_{j,\lambda}$ of each fluorescent entity as a modulating function, (2) the weighting of this function by the complex spectrum IT_λ, i.e.,

$$\Phi_{j,\lambda} = IT_\lambda \cdot I_{j,\lambda}. \tag{3}$$

The major advantages of this way of generating equation sets are (1) to maximize in each equation the relative weight of only one component (this procedure generates a system more diagonal than the unweighted one and therefore increases the value of the determinant); (2) to vary the weight of the data with respect to their intensity.

$$\sum_{\lambda=\lambda_0}^{\lambda_1} I_{1,\lambda} \cdot (IT_\lambda)^2 = a_1 \sum_{\lambda=\lambda_0}^{\lambda_1} (I_{1,\lambda})^2 \cdot IT_\lambda + \cdots + a_j \sum_{\lambda=\lambda_0}^{\lambda_1} I_{j,\lambda} \cdot IT_\lambda \cdot I_{1,\lambda} + \cdots$$
$$+ a_n \sum_{\lambda=\lambda_0}^{\lambda_1} I_{n,\lambda} \cdot IT_\lambda \cdot I_{1,\lambda}$$

$$\sum_{\lambda=\lambda_0}^{\lambda_1} I_{j,\lambda}\cdot(IT_\lambda)^2 = a_1 \sum_{\lambda=\lambda_0}^{\lambda_1} I_{1,\lambda}\cdot IT_\lambda\cdot I_{j,\lambda} + \cdots + a_j \sum_{\lambda=\lambda_0}^{\lambda_1} (I_{j,\lambda})^2\cdot IT_\lambda + \cdots$$

$$+ a_n \sum_{\lambda=\lambda_0}^{\lambda_1} I_{n,\lambda}\cdot IT_\lambda\cdot I_{j,\lambda}$$

$$\sum_{\lambda=\lambda_0}^{\lambda_1} I_{n,\lambda}\cdot(IT_\lambda)^2 = a_1 \sum_{\lambda=\lambda_0}^{\lambda_1} I_{1,\lambda}\cdot IT_\lambda\cdot I_{n,\lambda} + \cdots + a_j \sum_{\lambda=\lambda_0}^{\lambda_1} I_{j,\lambda}\cdot IT_\lambda\cdot I_{n,\lambda} + \cdots$$

$$+ a_n \sum_{\lambda=\lambda_0}^{\lambda_1} (I_{n,\lambda})^2 IT_\lambda.$$

This final equation set is then resolved using a classical routine. Once the results are obtained (i.e., values for $a_1, a_2, \ldots, a_n$), residues are calculated as the difference between a reconstructed complex spectrum and the experimental one. Weighted residues (WR) are then calculated as the ratio of the residues versus the absolute value of the random noise superposed to the signal. The fit between the result of the resolution and the experimental spectrum is represented by a graphic plot of the WR and a numerical estimator is calculated as the chi-square of the WR. For an optimally resolved spectrum, WR should be then randomly distributed about the zero value and the chi-square should be close to 1.

This technique allows one to calculate optimal contributions of several fluorescence spectra to the complex spectrum resulting of a mixture of several chemicals. This implies that the characteristic fluorescence spectrum of each fluorescent compound is introduced in the resolution system. The user must then record the characteristic spectrum of each form and store it in a library of spectra prior to analysis.

Some examples of application

a) Informations about the enzyme cellular microenvironment: the Calcium Probe. Because many enzymatic processes are energy-dependent, it may be of interest to check the ionic concentrations in their environment. Different cellular ionic fluorescent probes have been designed which allow the concentration of H_+, Mg^{2+} or Ca^{2+}. One of the most widely used probes for this last ion is 1H-indole-6-carboxylic acid, 2-{-4-(bis-carboxymethyl)amino-3-[-2(bis-carboxymethyl)amino-5-methylphenoxyethoxy]phenyl}, called Indo-1.

Preliminary studies of the properties of this probe in solution have demonstrated that, besides its acidobasic properties, Indo-1 was able to bind with proteins such as BSA to give a reversible 1/1 complex (Bancel 1992a). All these interactions having been proven to be exclusive to each other the scheme presented in Figure 20A is representative of the potential reactivity of Indo-1 in a biological environment (Bancel 1992b).

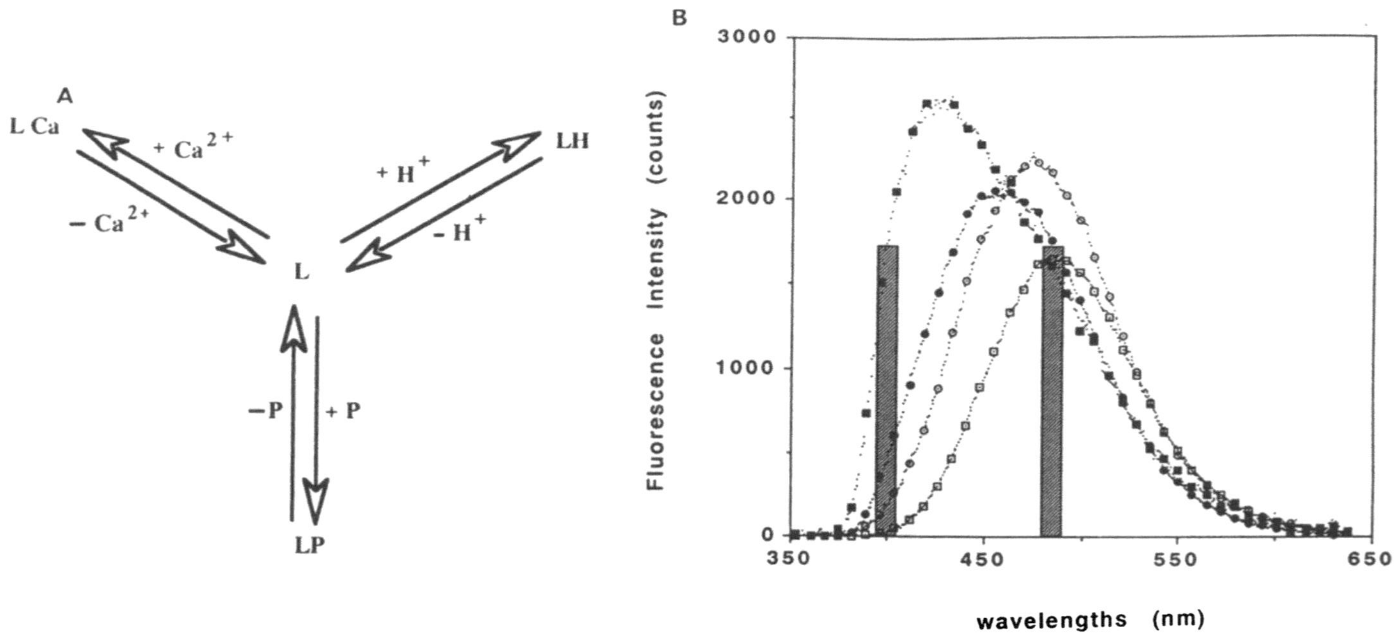

Fig. 20. **A** Schematic representation of the intracellular interactions of Indo 1 (*L*). **B** Characteristic fluorescence spectrum of each of the different chemicals involved in these equilibria: LM (■); L (□); LP (●); LH (○). The *gray bars* show the respective position of the filters used for the "ratio method" of determining Ca^{2+}

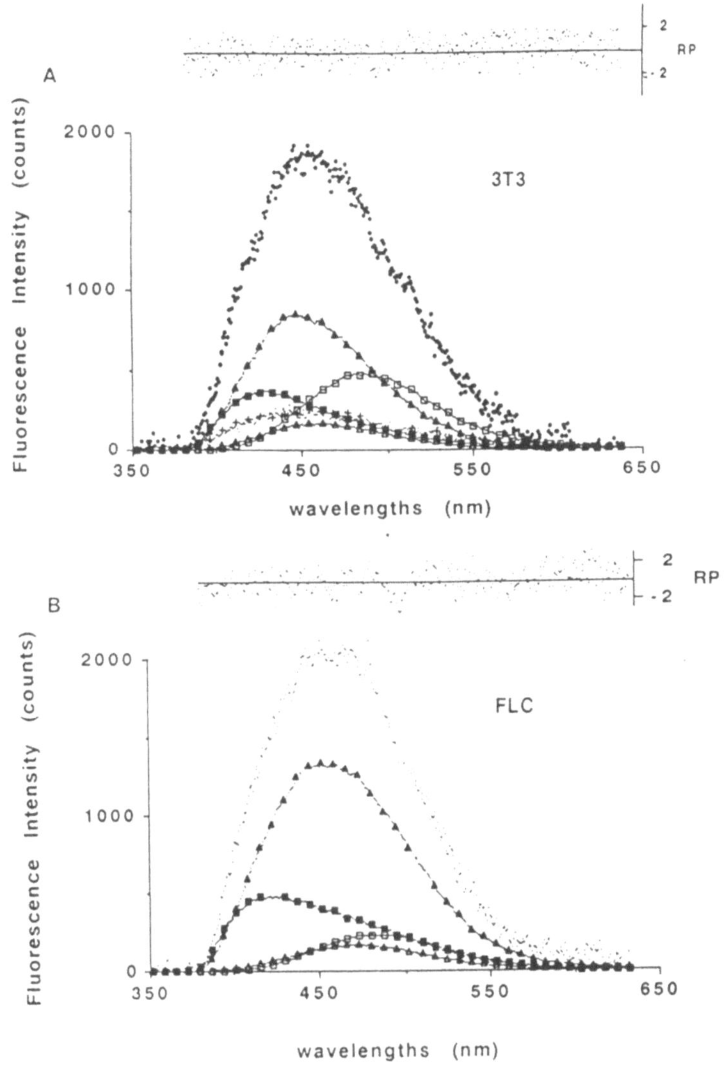

Fig. 21 A, B. Resolution of the cell fluorescence spectrum of: **A** 3T3 cell pre-incubated with Indo 1. The respective contribution of each participant was: LM (■) 17%; L (□) 23.4%, LP (▲) 39.4%; LH (△) 7.7%; intrinsic cell fluorescence spectrum (+) 14.2%. **B** A FLC cell incubated with Indo 1. The respective contribution of each participant was: LM (■) 22.3%; L (□) 9.8%; LP (▲) 60.3%; LH (△) 7.3%, which gave a pH of 6.78 and an intracellular Ca^{2+} concentration of 112 nM

Fortunately, each of the chemicals involved in these equilibria can be characterized by its own fluorescence spectra (Figure 20B). With this library of fluorescence spectra available, the resolution of the fluorescence spectrum of cells previously incubated with Indo-1 can be performed. Whatever the cell

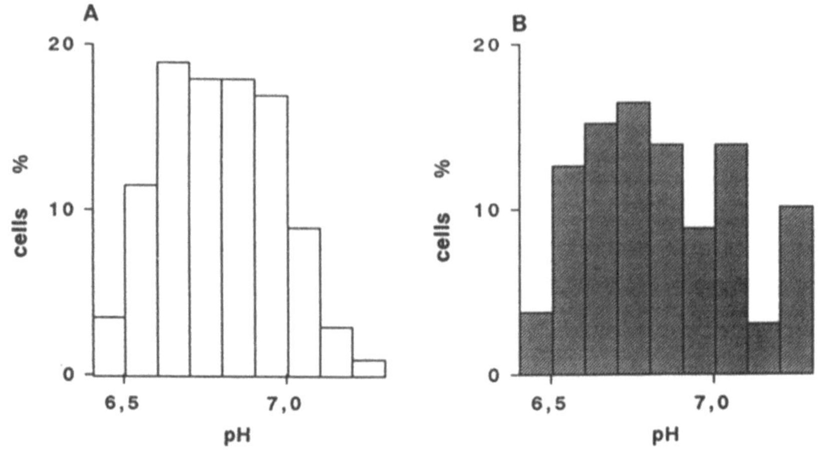

Fig. 22A, B. Indo 1 as an intracellular pH probe: comparison between results obtained with Indo 1 (**B**) and those obtained with the conventional pH probe DCH (dicyanobenzoquinone) (**A**); cells: 3T3 fibroblasts

types considered, each participant in the complex equilibria has been identified and quantified in the cells (Figure 21). Such results are consistent with numerous reports of anomalous intracellular spectra of this family of calcium probes. Furthermore, they demonstrate that:

1. Indo-1 may be used to probe the pH and the concentration of the ion calcium in a living cell (Figure 22). This has been verified in experiments where intracellular pH values of 3T3 cells were calculated from data separately obtained with the conventional pH probe DCH or with Indo-1 (Bancel 1992c).
2. For some cell types, the interactions with the cellular proteins are so important that they can account for more than 60% of the whole cell fluorescence (Figure 21B). This means that, in these cases, 60% of the recorded signal is of no use for the determination of the ionic concentrations.
3. The general use of a two-wavelength ratio has to be questioned at least for this application because (a) more than two chemicals participate in the whole system and (b) pH variations are said to be associated with variations in the Ca^{2+} concentration.

Finally, such an example emphasizes the need for careful studies of the chemical properties of any potential probe before its commercialization. It is not clear if the variations in the relative importance of interactions between Indo-1 and cellular proteins may result in significant and original information or if they only parallel the intracellular pH variations. What is clear, in contrast, is that up to 60% of the recorded signal may be due to unexpected and undesired interactions; such a situation is difficult to approve as so much care has to be taken to avoid any excess of probe in this kind of experiment.

b) Quantification of intermediate fluorescent metabolites. The metabolism of the polycyclic aromatic hydrocarbons (PAH) involves a very complex system of enzymes which has been extensively studied on purified enzymes and microsomal fractions. However, in such studies the balance between activating enzymes (i.e., aryl hydrocarbon hydroxylase) and detoxifying enzymes (i.e., transferases and epoxide hydrolase) is disturbed because of loss of membrane and cell integrity and lack of cytosolic cofactors and enzymes. Nevertheless, such studies have supplied essential information about the nature of intermediate metabolites. Most of these metabolites have been extensively studied, as they were suspected to be parents of potent carcinogens if not themselves carcinogenic. Most of them are available from the NIH repository, which allows one to build up a library of fluorescence spectra for most of the carcinogenic PAH and their metabolites, Finally, it is a very special situation in which probes can act as both witness and actor in the enzyme process.

Benzo(a)pyrene (B(a)P) is both a potent carcinogen and a fluorescent chemical. Its well-structured fluorescence spectrum can be easily obtained with excitation wavelengths which are not extensively absorbed by the optics of conventional microscopes. Thus, it has all the qualities required to be selected as a probe for intracellular studies of the metabolism of PAH.

As has been said before, intracellular accumulation of intermediate metabolites results from an existing imbalance between the processes of production and the processes of elimination of these metabolites. In this kind of experiment, the imbalance is created by preincubating the cells with a known concentration of substrate. The resolution of cell fluorescence spectra allows the identification of the accumulated metabolites, as is illustrated in Figure 23.

In this precise case, 9-hydroxybenzo(a)pyrene (9-OH-B(a)P) and 3-hydroxybenzo(a)pyrene (3-OH-B(a)P) have been identified by their respective characteristic fluorescence spectrum. Then successive records of cell fluorescence spectra allow to simultaneously monitor the evolution with time of the respective participation of each compound (i.e., B(a)P, 9-OH-B(a)P and 3-OH-B(a)P). From diagrams such as Figure 23B, the rate constant of the different reactions involved can be calculated. Results are the presented as histograms which display the distribution of the cell population as a function of rate constants (Figure 24).

When the second step (i.e., the detoxification step) of the metabolism is specifically studied, (Figure 25) the phenolic compounds 9-OH-B(a)P or 3-OH-B(a)P can be used as substrates (Anthelme 1990; Lautier 1988, 1990).

2.2.2. Correlation with biochemical conventional data

In a recent report, E Kohen (1989) wrote that the NAD(P)H transients recorded upon microinjection of bioenergetic substrates can often be modeled by the following equation:

$$F(t) = \frac{K_1}{K_1 - K_2} [S_0] \left(e^{K_2 t} - e^{K_1 t}\right),$$

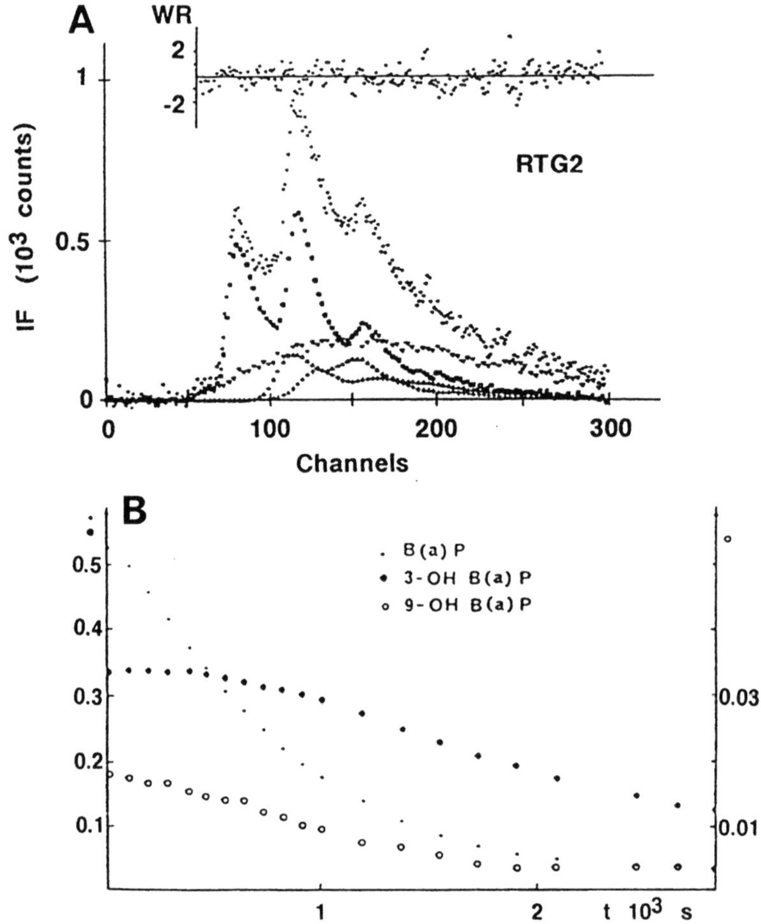

Fig. 23. A Intracellular accumulation of 9OH-B(a)P and 3OH-B(a)P during the metabolism of B(a)P. (●) experimental cell fluorescence spectrum; (□) B(a)P, (+) 3-OH-B(a)P, (▲) 9-OH-B(a)P respective fluorescence spectrum; (▼) intrinsic cell fluorescence. **B** Variation with time of the respective intracellular concentration of B(a)P, 3-OH-B(a)P and 9-OH-B(a)P in RTG2 cells

where F(t) is the intensity of NAD(P)H fluorescence, K_1 is the rate constant of its production, K_2 the rate constant of its reoxidation, $[S_0]$ is the initial concentration of the injected metabolite, and t is time. The fact that such an expression fits most of the results is hardly surprising, for all the steps involved in the process, either the diffusion processes of reactants and products or the enzyme reaction itself, can be expressed in terms of exponential functions. Moreover, as long as the reaction takes place in the cytoplasm, it can be expected from Udenfriend's experiment (1966) that the rate-determining step

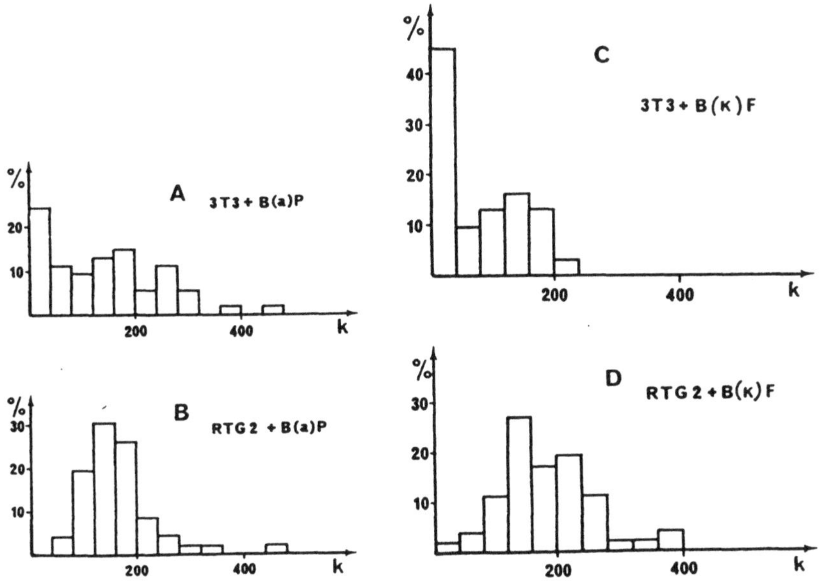

Fig. 24A–D. Frequency distribution of metabolic rate constants k of different metabolites in different cell lines. *B(a)P*benzo(a) pyrene; *B(k)F*benzo(k)fluoranthene

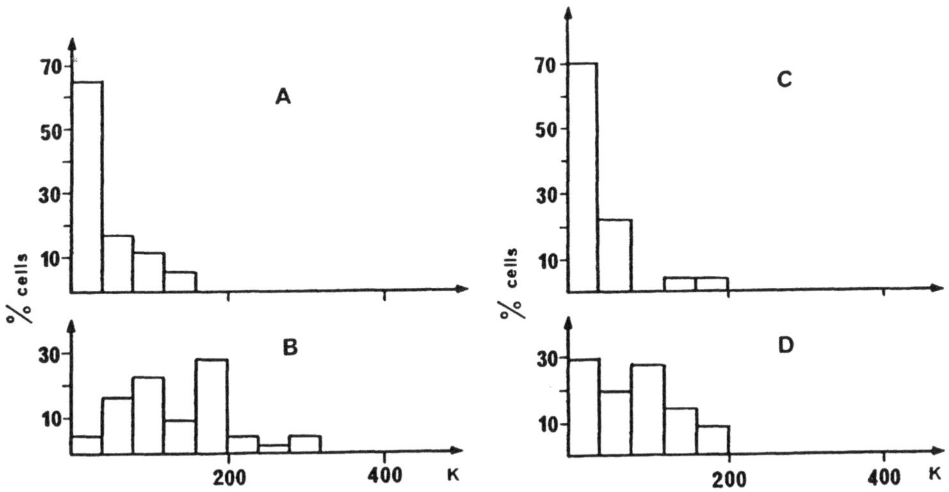

Fig. 25A–D. Frequency distribution of metabolic rate constants k, expressed in $10^{-5}S^{-1}$, of 3-OH-B(a)P in RTG2 cells **(A, B)**, or of 9-OH-B(a)P in 3T3 cells **(C, D)**. **A** and **C** Cells were preincubated in culture medium containing D-galactosamine 5 mM. **B** and **D** Cells were pre-incubated in normal culture medium. D-galactosamine is said to block the UDPG pathway, which is supposed to be the predominant one in RTG2 cells. Such results demonstrate that it is quite efficient in 3T3 cells

will be the enzyme reaction itself. The plateau quoted by E. Kohen (1989) in the NAD(P)H transient after microinjection of substrates like malate may reflect the fact that the rate of the whole process is decreased by some transmembranous transportation.

For such transient studies, the knowledge of the possible variations in free/bound NAD(P)H ratio (Vigo 1987) would be invaluable: no variation will indicate that the availability of NAD(P)H is not a limiting factor: on the contrary, any variation could be discussed in terms of excess of injected substrates resulting in a severely outbalanced system or of pathological evidence.

The same kind of exponential functions has been used to fit experimental results on PAH metabolism (Salmon 1980). Monoexponential decays have been observed for biotransformation of PAH by Cyt P450 and the metabolism of intermediate metabolites is well fitted by biexponential equations (cf. Figure 23B).

In all the above cases, discussions and interpretations will be made easier if the experimental rate constants can be expressed in terms more familiar to enzymologists, i.e., V_m, K_m, and eventually inhibition constants.

A first trial in this method was done by D. Lautier (1987) in order to discuss and explain the cell-to-cell variability of the rate constant she had observed in her study of the metabolism of benzo(a)pyrene. Her formalism can be easily generalized as follows.

Generally speaking, each step of an enzyme process can be described by the well-known chemical reaction

$$S_e + S_i + E \underset{k_{-1}}{\overset{k_1}{\rightleftarrows}} ES_e + S_i \underset{k_{-2}}{\overset{k_2}{\rightleftarrows}} ES_eS_i \overset{k_3}{\rightarrow} P + E,$$

in which S_e stands for the exogenous substrate, a fluorescent PAH for instance, S_i represents the intracellular substrate or cofactor, E is the enzyme, and ES_e and ES_eS_i the respective intermediate enzyme-substrate(s) complexes.

If v, the rate of the reaction, is expressed in terms of the intracellular concentration of S_e, i.e., $v = dS_e/dt$, then simple calculations result in the following expression:

$$k_{exp} = \frac{V_m}{K_m} = \frac{k_1 k_2 k_3 [E_T][S_i]}{k_{-1}(k_{-2} + k_3) + k_2 k_3 [S_i]},$$

where k_{exp} is the rate constant obtained for an experimental monoexponential decay, and E_T is the total intracellular concentration of the enzyme.

At this point it is important to recall that neither the intracellular concentration of E_T nor the available intracellular concentration of S_i is known. Nevertheless, a crude inspection of the above equation leads to the following remarks:

- at least for cytosolic enzymes, no variations have to be expected in the respective values of $k_1, k_2, \ldots$ when results of cell experiments are compared to

results obtained in solution. The above equation offers a way of linking conventional studies of enzyme activity to experiments in single living cells.
- because k_{exp} is dependent on the concentration of both E_T and S_i, the cell-to-cell variability of K_{exp} will reflect both the cell-to-cell variability in E_T concentration and the cell-to-cell variability of availability of S_i. It is thus not surprising to find relatively important differences in results from cells belonging to the same population. The histograms which represent the results are far more informative than a mean value.
- due to the fact that E_T and S_i do not interfere in the same way on the value of k_{exp}, this equation makes it possible to differentiate compounds which directly affect the amount of available enzyme (inhibitors) from those which act on the process by decreasing the amount of available S_i.

This method of data treatment has been applied to studies of the metabolism of B(a)P in cells, as different from 3T3 fibroblasts and RTG2 (Lahmy 1984, 1988a, b). The influence of different substrates on the glucuronide pathway has also been investigated (Lautier 1988, 1990; Anthelme 1990).

There is no doubt that the same kind of treatment could be applied to E. Kohen's experiments on the glycolytic pathway, but its optimal use will require fluorescent substrates of the different steps to be injected in order to monitor the changes with time of the concentrations of both substrates.

3. What next?

Trying to fix the next trends in science development is always hazardous. Nevertheless, a crystal ball does not seem necessary to predict that, whatever the efficiency of the techniques available today, scientists involved in the study of enzymes in single living cells will try in the near future to improve their knowledge by taking advantage of the following potentialities.

Spectral resolution in videomicrofluorometry

As has been stated, correct use of the videomicrofluorometric techniques makes it possible to obtain quantitative information on the whole microscopic field. If an image can be recorded at any specific wavelength and if there are i participants in the fluorescence, equation (1) of Section 2.2 is valid for each pixel of the image plane of the memory.

$$IT_\lambda = \sum_{i=1}^{N} a_i I_{i,\lambda} . \qquad (1)$$

If, then, there is enough room in memory, you can dream about storing enough images to be able to repeat the whole process of the numerical resolution of fluorescence spectra on each pixel of the image. Of course, it could be a good idea to minimize the size of such calculations by selecting the most informative wavelengths. This would also minimize the time necessary to record the information; another way to save time in collecting the data is to adapt the spectral width of the filters in order to collect the highest number of photons without losing the selectivity needed to solve the system.

Such an optimization of the data recording can be performed by inspection of the fluorescence spectrum of each participant in the fluorescent image. However, some sophisticated statistical methods can be used which will prove useful to avoid any involuntary bias of the data interpretation. It seems that this kind of application is on its way for some simple biological systems (Vigo 1993).

Scanning microfluorometry

In traditional stage scanning fluorometry, stepping stages are used that, under control of a computer, make small steps in the x and y directions. Although the number of steps per second can be quite high, it has been claimed that speeds of 100 steps per second are a practical limit (Schipper 1979). In our opinion, such high speeds are only available for qualitative observations, but significantly lower speeds have to be used if quantitative results are at stake. With a SIT camera, the scanning of the 512 channels needs 64 μs, but the time lag between two scans is 32 ms. It has been said before that 10 to 100 scan accumulations are often needed to reach a sufficient signal-to-noise ratio. CCD cameras are said to work at 25 images per second. The sensitivity of intensified cameras is of the same order of magnitude as that of the SIT camera, so that the time necessary to collect convenient data remains in the same range. With such detectors, data usable for further treatment cannot be obtained at a speed greater than ten steps per second. The above does not imply that fast scannings are unnecessary, because any loss of time between two successive data recordings has to be avoided.

In laser scanning microscopy, the object is not illuminated in total, but scanned step by step. The fluorescence light emitted by each illuminated point is recorded and an image is built up by storing these point-by-point measurements using usual analog-to-digital conversion. Three main types of laser scanning microscope are now available. Ultrafast ones using a rotating polygon will not be considered here for reasons previously indicated; but those based on laser beam mirror scanning (White 1987a, b) and stage scanning (Brakenoff 1985) have interesting properties for quantitative applications.

When E. Kohen et al. designed their 512·512 equipment, they were on the verge of designing a scanning microspectrofluorometer. Reducing in one direction the size of the slit which limits the size of the fluorescent beam, and using a diffraction grating with the dispersion direction perpendicular to the long direction of the slit, allowed them to simultaneously record the fluorescence spectra which theoretically originate from each of the 512 points imaged on the 512 lines of the camera. If the stage of the microscope had been equiped with a 0.5 – 1 μM stepping motor, they would have been able to record such spectra for each step of the motor. That would have given us a map of the cell which would display a quantitative distribution of each participant to the cell fluorescence spectrum if the signal-to-noise ratio was large enough to allow the use of a method of resolution of fluorescence spectrum. Both the cost of data storage and the slowness of data treatments preclude such developments at that time, especially for kinetics applications. On the other hand, the use of a conventional source would certainly have limited the selectivity of the step-to-step process of topographic selection. The situation is now completely different, and such an idea is being developed in some laboratories (R. Santus, pers. comm.).

The same ideas can be used to record fluorescence decays. In this case the 512 lines of the camera can be used to display the simultaneous decay with the time of the fluorescence intensity issuing from 512 small parts of a cell illuminated by a high-repetition-rate tunable laser (Vigo, pers. comm.).

In both cases, the use of powerful data processing implies that data are collected with a good signal-to-noise ratio. The true limits to such applications result from the amount of fluorescence which can be expected to reach each pixel of the camera. Because it has been previously claimed that low levels in irradiation flux and probe concentration have to be used, progress can only be achieved by using a very sensitive camera and minimizing the losses of light between the stage of the microscope and the front face of the detector.

3-D topographic mode using confocal microscope

A confocal fluorescence microscope is characterized by its ability to record information issuing from a well-defined volume of the object. The principle is shown in Figure 26.

A point source, irradiated with a laser, is imaged into a fluorescent object with a high-numerical-aperture objective. The same objective is used to re-image the image of the point source onto a pixel of the detector. Source and detection pinholes, with size lower than 10 μM, are optically separated by using a beam dichroic splitter.

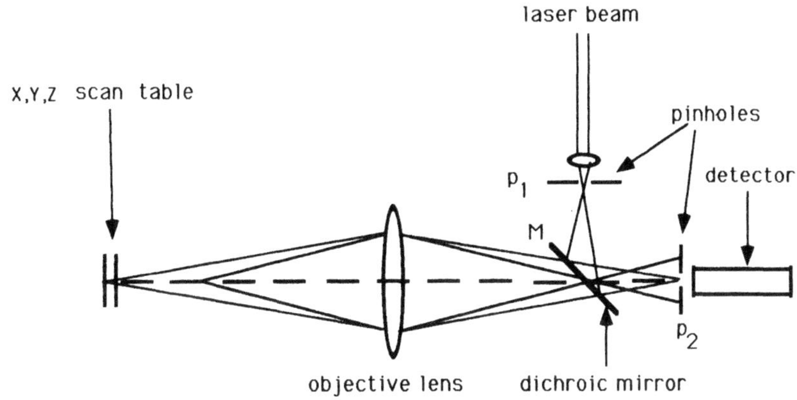

Fig. 26. Principle of the three-dimensional resolution of confocal microscopes

In this way, the total system response is given by the product of the excitation and detection response functions convoluted with the spatial object fluorescent distribution. The result is a real resolution along the optical axis and a slightly better resolution in the perpendicular plane than that obtained with the conventional microscope. From inspection of Figure 26, it is obvious that although all fluorescent molecules in the path of the excitation beam will be excited, only those located within the focal volume will contribute to the detector signal.

By successive computor-monitored movements of the microscope stage in the X, Y, Z directions, optical sectioning through the object is realized. The collected images can then be used for three-dimensional reconstructions of the object.

To our knowledge, confocal microscopes have not yet been used for intracellular biochemistry studies. Their wonderful performance has been applied to the study of structural modifications in organelles and cell membranes mainly because, as has been stated above, a strict data quantification is not required for such purposes. Nevertheless, some applications have to be expected in the near future if new probes, more specific in terms of compartimentalization, can be synthesized.

New trends in probe design

Recent improvements in the technologies available for studying the biology of living cells have opened the way to studies of events occurring at the organelle level. It has been demonstrated that both direct information on the kinetics of the enzyme activity, as well as information about modifications in their microenvironment, such as ionic concentrations can be obtained. Scientists may consider recording information on the kinetics of enzyme activities specifically located inside the mitochondria or embedded in the endoplasmic reticulum. For such a goal they do not necessarily need information on what occurs in other parts of the cell – though it could sometimes be of interest. Let us suppose that they are interested in the behavior of an enzyme located on the surface of the endoplasmic reticulum. The activity of this enzyme may be dependent on pH variations in its microenvironment which are not necessarily the same as those occuring in the whole cytoplasm. With the probes now available, only these last variations can be obtained. Likewise, the specific enzyme that they are interested in may exist also in other cell membranes. Unspecific information which describes the behavior of the whole cellular stock of this enzyme may be unrepresentative of the precise information that they are seeking.

A good example of a potential lack of specificity in the results obtained using fluorescence on single living cells is given by experiments using reduced nucleotides as fluorescent probes. Because both their fluorescence spectrum lifetimes are identical, variations in NADH and/or NADPH cannot be distinguished. Thus, activities of enzymes that use one of these as cofactor cannot be studied if no specific substrate can be injected into the cells.

Videomicrofluorometric techniques are the most efficient and easy to use when the correction image is as dark as possible. Apart from the fluorescence due to the optics of the equipment, partial illumination of the microscopic field results from the fact that many fluorescent probes do not totally reach their target. Some molecules remain in the cytoplasm or find a niche in cellular membranes different from those in which they are supposed to be located. The image contrast is then reduced and the difficulties in segmenting the image increased. Segmentation is the operation by which, manually or automatically, the parts of the screen which contain useful information and which have to be stored for further treatment are selected. This permits both saving room in storage and speeding up the further treatment of data.

New probes must thus be designed which possess a double specificity. Scientists need probes specific to one characteristic of the enzyme microenvironment which can be specifically vectorized to a preliminary defined target. Such a goal will mean a lot of work for organic chemists and biochemists, who will be in charge first of selecting the right biological vector, and then of designing the convenient synthesis; but physicochemists will also have to carefully check the properties of the new drug. It can only be expected that, remembering some previous misadventures, these preliminary tests will be seriously performed before the drug is made commercially available.

4. General conclusion

As in many other fields of science, information in biology is acquired at dramatically different scales of resolution. Identification of the structures and functions of essential macromolecules is expected to help us in understanding the behavior of cells, group of cells, and living creatures. Depending upon the biological field in which recent progress has been made, the interest of scientists is focused mainly on molecular biology or on medical applications.

Between these extrema stands cellular biology. Living cells are the first level of complexity where the wonderful discoveries of molecular biology face the reality of life. Furthermore, recent moral considerations have resulted in restrictions on the use of animals in the laboratory, which increases the potential of experiments on intact living cells as an intermediate stage between purely *in vitro* and *in vivo* experiments.

Astonishingly fast progress in genetics has opened the way to countless potential applications in medicine and enzyme engineering. Genetically modified cells have already been created, and many more are expected, which might be used to cure human enzyme deficiency or to synthesize pharmaceutical or industrial molecules. The first step in such a direction is to verify on intact living material if the yield of the modified enzymatic system is the one expected from *in vitro* studies. Broad verifications can be done with fast methods such as flow cytometry, but detailed analyses of features require more sophisticated equipment.

Such a situation may be seen as favorable for the rapid development of quantitative studies on living cells. Positive results may be expected from the use of the equipment described in this chapter if some fundamental rules are obeyed and some limitations well assessed. Let us start with biological considerations: the dimensions of most of the cellular organelles are below the limits of resolution of optic spectroscopy. As a first consequence, whatever the selectivity of the probe and the sensitivity of the equipment, the recorded data will at best reflect the behavior of the pool of organelles located in the small part of the cell studied.

Secondly, the organelles of living cells have to be suspected of moving during the experiment. If such displacements are not randomized, they could result in successive increases and decreases in the number of organelles located in the cellular volume under study. This will induce fluctuations which may blur the kinetics of the enzyme process studied.

Finally, it can be expected that no synchronization exists in the functioning of the different organelles of a cell in the absence of external stimulus. Adding an excess of (fluorescent) substrate may induce some artificial synchronism in the functioning of these different organelles. To what extent this situation mimics the way in which cells react to specific stimuli remains to be proven. This might be quite a difficult task. To the above listing of the limits and difficulties that scientists will have to face, limitations specific to the fluorescence optic methods of investigation have to be added. Limitation in the local drug concentration, limitation in the flux of excitation light conflict with the necessity of recording data with low signal-to-noise ratio in time lags compatible with the speed of the enzyme process under study. The choice of the size of the element of volume on which the information will be recorded depends on all these biological and physical limitations, and such a choice, in turn, associated with other practical considerations, will lead to the use of a specific instrument and protocol. Thus, experimental protocols appear only as necessary provisional compromises, a situation which should temper some immodest claims.

Another matter of concern is the unconsidered use of concepts which have been defined at the macroscopic scale to characterize homogeneous systems, but which may lose most of their significance at the molecular level in inhomogeneous structured media. More often, the use of familiar but inadequate words only contributes to a misunderstanding or a misinterpretation of otherwise significant data.

References

Alibaud R, Salmon JM, Vigo J, Viallet P (1988) Utilisation du 1,4 Diacetoxi 2,3 dicyanobenzol (ADB) pour la détermination, par microspectrofluorimétrie, du pH intracellulaire de cellules vivantes isolées. *C. R. Acad. Sci.* (Paris) 307:89–92

Allegre JM, Salmon JM, Commalonga J, Savelli M, Viallet P (1985) Microspectrofluorimétrie quantitative: automatisation d'un microspectrofluorimère équipé d'une platine motorisée permettant l'enregistrement simultané, sur différentes cellules d'une population cellulaire, de cinétiques intracellulaires. *Innov. Tech. Biol. Med.* 6:743–752

Alsins J, Claesson S, Elmgen H (1982) A simple instrumentation for measuring fluorescence lifetimes of probe molecules in small systems. *Chem. Scr.* 20:183

Ameloot M, Hendrickx (1982) Criteria for model evaluation in the case of deconvolution calculations. *J. Chem. Phys.* 76:4419

André JC, Bouchy M, Donner M (1987) On diffusion in organized assemblies and in biological membranes. *Biorheology* 24:237–272

Anthelme B, Lautier D, Salmon JM, Vigo J, Viallet P (1990) Influence of controlled glucose deprivation on kinetics of detoxification mechanism for two intermediate metabolites, 9-hydroxyben-zo(a)pyrene and 3-hydroxybenzo(a)pyrene in 3T3 and RTG2 Cells. *Polycyclic Aromatic Compounds* 1:71–79

Atherton SJ, Beaumount PC, (1984) Ethidium bromide as a fluorescent probe of the accessibility of water to the interior of DNA. *Photochem. Photobiophys.* 8:103

Bancel F, Salmon JM, Vigo J, Vo-Dinh T, Viallet P (1992a) Investigation of noncalcium interactions of Fura-2 by classical and synchronous fluorescence spectroscopy. *Anal. Biochem.* 204:231–238

Bancel F, Salmon JM, Vigo J, Viallet P (1992b) Microspectrofluorometry as a tool for investigation of non-calcium interactions of indo-1. *Cell Calcium* 13:59–68

Bancel F (1992c) Approche microspectrofluorimetrique et physicochimique des sondes fluorescentes à calcium: Application à la mesure simultanée du calcium et du pH intracellulaires sur cellules vivantes isolées. Thesis, University of Perpignan, Perpignan, France

Basu S (1967) Ultraviolet absorption studies on DNA. *Biopolymers* 16:2315–2328

Benson DM, Bryan J, Plant AL, Gotto AM Jr, Smith LC (1985) Digital imaging fluorescence microscopy: spatial heterogeneity of photobleaching rate constants in individual cells. *J. Cell. Biol.* 100:1309–1323

Berlin RD, Fera P (1977) Changes in microviscosity associated with phagocytosis: effects of colchicine. *Proc. Natl. Acad. Sci. USA* 74:1072–1076

Blais J, Amirand C, Nocentini S, Ballini JP, Vigny P (1992) Photoreactions of furocoumarins in human fibroblasts: a microspectrofluorometric study. *J. Cell. Pharmacol.* 3:151–156

Bottiroli G, Prenna G, Andreoni A, Sacchi CA, Svelto O (1979) Fluorescence of complexes of quinacrine mustard with DNA. Influence of the DNA base composition on the decay time in bacteria. *Photochem. Photobiol.* 29:23–28

Bouchy M, Donner M, André JC (1991) Erythrocyte membranes alteration in a shear stress measured by fluorescence anisotropy. *Cell Biophysics* 17:213–225

Brakenoff GJ, van der Voort HTM, van Sprouseb EA, Linnemans WAM, Nanninga N (1985) Three-dimensional chromatin distribution in neuroblastoma nuclei shown by confocal scanning laser microscopy. *Nature* (London) 317:748–749

Brochon JC, Wahl P (1972) Mesure des declins de l'anisotropie de fluorescence de la γ globuline et de ses fragments Fab, Fc et F(ab)2 marqués avec le 1-sulfonyl-5-dimethylaminonaphthalène. *Eur. J. Biochem.* 25:20–32

Brochon JC, Wahl P, Jallon JM, Iwatsubo M (1976) Pulse fluorimetry study of beef liver glutamate dehydrogenase-reduced nicotinamide adenine dinucleotide phosphate complexes. *Biochemistry* 15:3259

Chiu HC, Bersohn R (1977) Electronic energy transfer between tyrosine and tryptophan in the peptides Trp-(Pro)-Tyr. *Biopolymers* 16:277–288

Choi HS, Dilley R, Kim Y, Schwartz SM (1988) Proceedings of the 10th Annual International Conference of the IEEE Engineering in Medicine & Biology Society 10:261–362

Chused TM (1989) Flow cytometric measurements of physiologic cell responses. In: *Cell structure and function by microspectrofluorometry* Kohen E, Hirschberg JG eds pp. 377–389, Academic Press, San Diego

Collins JM, McLean Grogan W (1989) Comparison between flow cytometry and fluorometry for the kinetic measurement of membrane fluidity parameters. *Cytometry* 10:44–49

Dale RE, Eisinger J (1974) Intramolecular distances determined by energy transfer. Dependance on orientional freedom of donor and acceptor. *Biopolymers* 13:1573

Dale RE (1978) Fluorescence depolarization and orientation factors for excitation energy transfer between isolated donor and acceptor fluorophore pairs at fixed intermolecular separation. *Acta. Phys. Polon* A54:743–756

Donner M, Stoltz JF (1985) Comparative study on fluorescent probes distributed in human erythrocytes and platelets. *Biorheology* 22:385–397

Donner M, Muller S, Stoltz JF (1989) Molecular rheology of white blood cells. *Rev. Port. Hemorhologia* 3:235–247

Donner M, Muller S, Stoltz JF (1990) Fluorescence depolarization method in the study of dynamic properties of blood cells. *Biorheology* 27:367–374

Duportail G, Mauss Y, Chambron J (1977) Quantum yields and fluorescence lifetimes of acridine derivatives interactions with DNA. *Biopolymers* 16:1397–1404

Eisinger J, Blumberg WE, Dale RE (1981) Orientational effects in intra- and intermolecular long range excitation energy transfer. *Ann. N. Y.* Acad. Sci 366:155

Eyl M, Muller S, Donner M, Maugras M, Stoltz JF (1992) Use of fluorescence anisotropy determinations for indicating the physiological status of hybridoma cell cultures. *Cytotechnology* 8:5–11

Fairclough RH and Cantor CR (1978) The use of singlet-singlet energy transfer to study macromolecular assemblies. *Methods Enzymol.* 48:347–379

Finney DA, Sklar LA (1983) Ligand receptor internalization: a kinetic, flow cytometric analysis of the internalization of N-formyl peptides by human neutrophils. *Cytometry* 4:54–60

Förster Th (1948) Intermolecular enery transference and fluorescence. *Ann Phys* (*Leipzig*) 2:55

Förster Th (1949) Experimental and theoretical investigation of intermolecular transfer of electron activation energy. *Z Naturforsch* 4a:321–7

Gafni A, Brand L (1976) Fluorescence decay studies of reduced nicotinamide adenide dinucleotide in solution and bound to liver alcohol deshydrogenase *Biochemistry* 15:3165

Gratton E, Jameson DM, Hall RD (1984) Multifrequency phase and modulation fluorometry. *Annu. Rev. Biophys. Bioeng.* 13:105–124

Hare F, Lussan C, Sanchez E (1976) Apparatus for the direct measurement of fluorescent decay as applied to biological membranes and their various models. *J. Chim. Phys-Chim, Biol.* 73:621

Hause LL, Clowry LJ, Megan PA (1988) Microscopic image analysis in cervical cytology. In: *Proceedings of the 10th Annual International Conference of the IEEE Engineering in Medicine & Biology Society* 10:375–376

Inbar M, Shinitzky M (1975) Decrease in microviscosity of Iymphocyte surface membrane associated with stimulation induced by Concanavaline A. *Eur. J. Immunol.* 5:166–170

Inbar M, Larnicol N, Jasmin C, Mishal Z, Augery Y, Rosenfeld C, Mathé G (1977) A method for the quantitative detection of human acute Iymphatic leukemia. *Eur. J. Can.* 13:1231–1236

Irvin JA, Quickenden TI, Sangster DF (1981) Criterion of goodness of fit for deconvolution calculations. *Rev. Sci. Instrum.* 52:131

Jasmin C, Augery Y, Calvo F, Iarnicol N, Rosenfeld C, Mathé G, Inbar M (1981) Cellular blood fluorescence polarization: a possible prognostic tool in human acute lymphatic leukemia. *Biomedicine* 34:23–28

Johnson SM, Nicolau C (1977) The distribution of 1,6-diphenylhexatriene fluorescence in normal human lymphocytes. *Biochem. Biophys. Res. Commun.* 76:869–874

Kohen E (1964) Pyridine nucleotide compartmentalization in glass-grown ascites cells. *Exp. Cell. Res.* 35:303–316

Kohen E, Legallais V, Kohen C (1966) Introduction to microelectrophoresis and microinjection techniques in microfluorimetry. *Exp. Cell. Res.* 41:223–226

Kohen E, Kohen C (1974) Continuous helium-cadmium laser as an excitation source for the microspectrofluorometric assay of NAD(P)H in single living cells. *Int. J. Radiat. Biol.* 26:97–100

Kohen E, Kohen C, Hirschberg JG, Wouters AW, Bartick PR et al. (1981a) Examination of single cells by microspectrophotometry and microfluorometry. In: *Techniques in Life Sciences.* Baker PF ed. pp. 103–128, Elsevier/North-Holland, New York

Kohen E, Thorrel B, Hirschberg JG, Wouters AW, Bartick PR et al. (1981b) Microspectrofluorometric procedures and their applications in biological systems. In *Modern fluorescence spectroscopy*, vol 3 Wehry EL ed. pp. 296–346, Plenum, New York

Kohen E, Kohen C, Hirschberg JG, Wouters AW, Thorrel B et al. (1983) Metabolic control and compartmentation in single living cells. *Cell Biochem. Funct.* 1:13–16

Kohen E, Kohen C, Reyftmann JP, Morliere P, Santus R (1984) Microspectrofluorometry of fluorescent photoproducts in photosensitized cells. Relationship to cellular quiescence and senescence in culture. *Biochim. Biophys. Acta.* 805:332–336

Kohen E, Welch GR, Kohen C, Hirschberg JG, Bereiter-Hahn J (1986a) Experimental analysis of spatiotemporal organization of metabolism in intact cells. The enigma of "metabolic channeling" and "metabolic compartmentation" In: *The Organization of Cell Metabolism* Welch GK, Clegg JS eds. pp. 251–275, NATO Adv Research Workshop, Hanstholm, Denmark, Plenum, New York

Kohen E, Kohen C, Morliere P, Santus R, Reyftmann JP, Dubertret L et al. (1986b) A microspectrofluorometric study of the effect of anthralin, an antipsoriatic drug, on cellular structure and metabolism. *Cell. Biochem. Funct.* 4:157–168

Kohen E, Reyftmann JP, Morliere P, Santus R, Kohen C, Mangel WF et al. (1986c) A microspectrofluorometric study of porphirin-photosensitized single living cells. Part II: Metabolic alterations. *Photochem. Photobiol.* 44:471–475

Kohen E, Kohen C, Hirschberg JG, Santus R, Schachtschabel DO, Nestor J (1989) Microspectrofluorometry of single living cells: Quo Vadis. In: *Cell structure and function by microspectrofluorometry.* Kohen E, Hirschberg JG eds. pp. 199–228, Academic Press, San Diego

Kohen E, Kohen C, Prince J, Pinon R, Hirschberg JG, Santus R et al. (1992) Microspectrofluorometry of organelle interactions in hepatocytes treated with cytotoxic agents. *J. Cell. Pharmacol.* 3:8–21

Kurhy JG, Fonteneau P, Duportail G, Maechling C, Laustriat G (1983) TMA-DPH: a suitable fluorescence polarization probe for specific plasma membrane fluidity studies in intact living cells. *Cell. Biophys.* 5:129–140

Kuhry JG, Duportail G, Bronner C, Laustriat G (1985) Plasma membrane fluidity measurement on whole living cells by fluorescence anisotropy of trimethylammoniumdiphenylhexatriene. *Biochim. Biophys. Acta.* 845:60–67

Kutchai H, Huxley VH, Chandler LH (1982) Determination of fluorescence polarization of membranes probes in intact erythrocytes. *Biophys. J.* 39:229–232

Llenger D, Poindron P, Fonteneau P et al. (1990) The plasma membrane internalization and recycling in enhanced in macrophages upon activation with gamma-interferon and lipopolysaccharide; a study using the fluorescent probe trimethylaminodiphenylhexatriene. *Biochim. Biophys. Acta.* 1030:73–81

Lahmy S, Salmon JM, Viallet P (1988a) Influence of 3-Methylcholanthrene or carbaryl long term treatment on mixed function oxidase activity in 3T3 fibroblasts by single living cell investigation. *Cell Biochem. Funct.* 6:275–282

Lahmy S, Salmon JM, Viallet P (1988b) MFO activity changes in single 3T3 fibroblasts treated with Adriamycin. *Anticancer Res.* 8:1411–1418

Lahmy S, Salmon JM, Vigo J, Viallet P (1989) pHi and DNA content modifications after ADR treatment in 3T3 fibroblasts. A microfluorimetric approach. *Anticancer Res.* 9:929–936

Lakowicz (1989) Principles of frequency-domain fluorescence spectroscopy and applications to protein fluorescence. In *Cell structure and function by microspectrofluorometry.* Kohen E, Hirschberg JG eds. pp. 163–184, Academic Press, San Diego

Latt SA, Cheung HT, Blout ER (1965) Energy transfer. A system with relatively fixed donor-acceptor separation. *J. Am. Chem. Soc.* 87:995

Lautier D (1987) Application de la microspectrofluorimétrie à l'étude intracellulaire des oxydases à fonction mixte dans les cellules RTG2. Thesis, University of Montpellier 1, Montpellier, France

Lautier D, Salmon JM, Anthelme B, Viallet P (1988) 6-Aminochrysene, a v potent inhibitor of transferase activity in single living RTG2 cells. *J. Histochem. Cytochem.* 36:685–691

Lautier D, Anthelme B, Salmon JM, Vigo J, Viallet P (1990) Influence of d-galactosamine on the kinetics of metabolic processes for two intermediate metabolites, 9-hydroxybenzo(a)pyrene and 3-hydroxybenzo(a)pyrene in 3T3 and RTG2 cells. *J. Histochem. Cytochem.* 38:949–963

Maire C, Bouchy M, Donner M, André JC (1993) Membrane labelling by fluorescent probes: incorporation of TMA-DPH in erythrocyte membranes. *Biorheology* (in press)

Malle E, Griess A, Kosntner GM, Pfieffer K, Nimpf J, Hermetter A (1989) Is there any correlation between platelet aggregation, plasma lipoproteins, apoproteins and membrane fluidity of human blood platelets? *Thromb. Res.* 53:181

Morelle B, Salmon JM, Vigo J, Viallet P (1993) Proton, Mg^{2+} and protein as competing ligands for the fluorescent probe, Mag-Indo-1, a first step to the quantification of intracellular Mg^{2+} concentration. *Photochem. Photobiol.* (in press)

Morliere P, Kohen E, Reyftmann JP, Santus R, Kohen C, Maziere JC (1987) Photosensitization by porphirins delivered to L cell fibroblasts by human serum low density lipoproteins. A microspectrofluorometric study. *Photochem. Photobiol.* 46:183–191

Muller CP, Volloch Z, Shinitzky M (1980) Correlation between cell density, membrane fluidity and the availability of transferring receptors in Friend erythro leukemic cells. *Cell. Biophys.* 2:233–240

Muller S, Donner M, Drouin P, Stoltz JP (1987) Lipid fluidity of erythrocyte membrane: failure to demonstrate significant alterations in diabetes mellitus. *Clinical Hemorheol.* 7:619–626

Muller S, Masson V, Droesch S, Donner M, Stoltz JF (1989) Application to the evaluation of opsonozing properties of fibronectin. *Biorheology* 26:323–330

Murphy RF, Jorgensen ED, Cantor CR (1982a) Kinetics of histone endocytosis in Chinese hamster ovary cells. A flow cytometric analysis. *J. Biol. Chem.* 257: 1695–1701

Murphy RF, Powers S, Verderame M, Cantor CR, Pollack R (1982b) Flow cytometric analysis of insulin binding and internalization by Swiss 3T3cells. *Cytometry* 2:402–406

Murphy RF, Powers S, Cantor CR (1984) Endosome pH measured in single cells by dual fluorescence flow cytometry: rapid acidification of insulin to pH 6. *J. Cell. Biol.* 98:1757–1762

Murphy RF, Roederer M (1986) In *Application of fluorescence in the biomedical sciences* Taylor DL, Waggonner AS, Murphy RF, Lanni F, Birge R eds. pp. 545–566, Alan R Liss, New York

Murphy RF (1989) Flow cytometric analysis of ligand binding and endocytosis. In *Cell structure and function by microspectrofluorometry* Kohen E, Hirschberg JG eds. pp. 363–376, Academic Press, San Diego

Plant AL, Benson DM, Smith LC (1985) Cellular uptake and intracellular localization of benzo(a)pyrene by digital fluorescence imaging microscopy. *J. Cell. Biol.* 100: 1295–1308

Prendergast FG, Haugland RP, Callahan PJ (1981) 1-4-(Trimethylamino)-phenyl-6-phenylhexa-1,3,5-triene: Sunthesis, fluorescence properties and use as a fluorescence probe of lipid bilayers. *Biochemistry* 20:7333–7338

Prenna G, Bottiroli G, Mazzini G (1977) Cytofluorimetric quantification of the activity and reaction kinetics of acid phosphatase. *Histochem. J.* 9:15–30

Reyftmann JP, Kohen E, Morliere P, Santus R, Kohen C, Mangel WF et al. (1986) A microspectrofluorometric study of porphyirin-photosensitized single living cells. 1. Membrane alterations. *Photochem. Photobiol.* 44:461–469

Salmon JM, Viallet P (1977) Use of electronic spectra in the study of enzymes-pyridine nucleotides interactions. *J. Chim. Phys., Phys-Chim., Biol.* 74:239–245

Salmon JM (1980) Réalisation d'un microspectrofluorimètre. Application à l'étude de quelques mécanismes cellulaires. Thesis, University of Perpignan, Perpignan, France

Salmon JM, Kohen E, Viallet P, Hirschberg JG, Wouters AW, Kohen C, Thorell B (1982) Microspectrofluorometric approach of the study of the free/bound NAD(P)H ratio as metabolic indicator in various cell types. *Photochem. Photobiol.* 36:525

Salmon JM, Vigo J, Viallet P (1988) Resolution of complex fluorescence spectra recorded on single unpigmented living cells using a computerized method. *Cytometry* 9:25–32

Santus R, Morliere P, E Kohen (1991) The photobiology of the living cell as studied by microspectrofluometric techniques. *Photochem. Photobiol.* 54:1071–1077

Scott TG, Spencer RP, Leonard J, Weber G (1970) Emission properties of NADH. Studies of fluorescence lifetimes and quantum efficiencies of NADH, AcPyADH and simplified synthetic models. *J. Am. Chem. Soc.* 92:687

Schipper J, Tilders FJH Ploem JS (1979) Extraneuronal catecholamine as an index for sympathetic activity: a scanning microfluorimetry study in the iris of the rat. *J. Pharmacol. Exp. Ther.* 211:265–270

Sipe DM, Murphy RF (1987) High-resolution kinetics of transferrin acidification in BALB/c 3T3 cells: exposure to pH 6 followed by temperature sensitive alkalinization during recycling. *Proc. Natl. Acad. Sci. USA* 84:7119–7123

Steinberg IZ (1978) Circular polarization of luminescence: biochemical and biophysical applications. *Annu. Rev. Biophys. Bioeng.* 7:113–137

Stoltz JF, Donner M (1985a) Relations between molecular rheology and red blood cell structure: methods and clinical approaches. *Clinical Hemorheol.* 5:813–848

Stoltz JF, Donner M (1985b) Fluorescence polarization applied to cellular microrheology. *Biorheology* 22:227–247
Stryer L, Haugland RP (1967) Energy transfer: a spectroscopic rules. *Proc. Natl. Acad. Sci. USA* 58:719
Teale FJW (1984) Phase and modulation fluorimetry. In Time resolved fluorescence spectroscopy Cundall RB, Dale RE eds. pp. 59–79, Plenum Press, New York
Thaer AA, Sernetz M (eds) (1973) Fluorescence techniques in cell biology. Springer, Berlin Heidelberg New York
Tsunoda Y, Yodozawa S, Tashiro Y (1988) Fluorescence digital image analysis of the inosital triphospate-mediated calcium transient in single permeabilized parietal cells. *FEBS Lett.* 231:29–35
Udenfriend S, Zaltzman-Nirenberg P, Guroff G (1966) A study of cellular transport with the fluorescent amino acid aminonaphthylamine. *Arch. Biochem. Biophys.* 116: 261–270
Valeur B, Moirez J (1973) Analyse des courbes de décroissance multiexponentielles par la méthode des fonctions modulatrices. Application à la fluorescence. *J. Chim. Phys.* 70:500–506
Valeur B (1978) Analysis of time-dependant fluorescence experiments by the method of modulating functions with special attention to pulse fluorometry. *Chem. Phys.* 3: 85–93
Valeur B (1993) Fluorescence probes for evaluation of local physical and structural parameters. In *Molecular luminescence spectroscopy. Methods and Applications: Part* 3 Schulman SG ed. Wiley-Interscience, New York
Vigo J, Salmon JM, Viallet P (1987) Quantitative microfluorometry of isolated living cells with pulsed excitation: development of an effective and relatively inexpensive instrument. *Rev. Sci. Instrum.* 58:1433–1438
Vigo J, Salmon JM, Lahmy S, Viallet P (1991) Fluorescent image cytometry: from qualitative to quantitative measurements. *Anal. Cell. Pathol.* 3:145–165
Wahl P (1965) Sur l'étude des solutions de macromolécule par la décroissance de fluorescence polarisée. *C. R. Acad. Sci.* (Paris) 260:6891–6893
Ware WR (1971) Transient luminescence measurements. In *Creation and detection of the excited state, vol* 1 *part A* Lamola AE (ed) Marcel Dekker, New York, pp 213–302
Ware WR (1983) Transient luminescence measurements. In *Time resolved fluorescence spectroscopy*, Cundall RB, Dale RE eds. pp. 23–58, Plenum Press, New York
Weber G (1952) polarization of the fluorescence of macomolecules. 1. Theory and experimental method. *Biochem. J.* 51:145–155
Weber G (1953) Rotational brownian motion and polarization of the fluorescence of solutions. In: *Adv. Protein. Chem.* 8:415–459
Weber G (1984) Old and new developments in fluorescence spectroscopy. In *Time resolved fluorescence spectroscopy* Cundall RB, Dale RE eds. pp. 1–20, Plenum Press, New York
Weber G (1989) From solution spectroscopy to image spectroscopy. In *Cell structure and function by microspectrofluorometry* Kohen E, Hirschberg JG eds. pp. 71–85, Academic Press, San Diego
White JG, Amos WB (1987a) Confocal microscopy comes of age. *Nature* (London) 328:183–184
White JG, Amos WB, Fordham M (1987b) An evaluation of confocal versus conventional imaging of biological structures by fluorescence light microscopy. *J. Cell. Biol.* 105:41–48

Printing: Druckhaus Beltz, Hemsbach
Binding: Buchbinderei Schäffer, Grünstadt

MIX
Papier aus verantwortungsvollen Quellen
Paper from responsible sources
FSC® C105338

If you have any concerns about our products,
you can contact us on
ProductSafety@springernature.com

In case Publisher is established outside the EU,
the EU authorized representative is:
Springer Nature Customer Service Center GmbH
Europaplatz 3, 69115 Heidelberg, Germany

Printed by Libri Plureos GmbH
in Hamburg, Germany